Quantum Electrodynamics

Advanced Book Classics
David Pines, Series Editor

Anderson, P.W., *Basic Notions of Condensed Matter Physics*
Bethe H. and Jackiw, R., *Intermediate Quantum Mechanics, Third Edition*
Feynman, R., *Photon-Hadron Interactions*
Feynman, R., *Quantum Electrodynamics*
Feynman, R., *Statistical Mechanics*
Feynman, R., *The Theory of Fundamental Processes*
Nozières, P., *Theory of Interacting Fermi Systems*
Pines, D., *The Many-Body Problem*
Quigg, C., *Gauge Theories of the Strong, Weak, and Electromagnetic Interactions*

QUANTUM ELECTRODYNAMICS

RICHARD P. FEYNMAN
late, California Institute of Technology

CRC Press
Taylor & Francis Group
Boca Raton London New York

CRC Press is an imprint of the
Taylor & Francis Group, an **informa** business

First published 1998 by Westview Press

Published 2018 by CRC Press
Taylor & Francis Group
6000 Broken Sound Parkway NW, Suite 300
Boca Raton, FL 33487-2742

ISBN 13: 978-0-201-36075-2 (pbk)
ISBN 13: 978-0-367-32008-9 (hbk)
ISBN 13: 978-0-429-49324-9 (ebk)

Visit the Taylor & Francis Web site at
http://www.taylorandfrancis.com

and the CRC Press Web site at
http://www.crcpress.com

Cover design by Suzanne Heiser

Editor's Foreword

Addison-Wesley's *Frontiers in Physics* series has, since 1961, made it possible for leading physicists to communicate in coherent fashion their views of recent developments in the most exciting and active fields of physics—without having to devote the time and energy required to prepare a formal review or monograph. Indeed, throughout its nearly forty-year existence, the series has emphasized informality in both style and content, as well as pedagogical clarity. Over time, it was expected that these informal accounts would be replaced by more formal counterparts—textbooks or monographs—as the cutting-edge topics they treated gradually became integrated into the body of physics knowledge and reader interest dwindled. However, this has not proven to be the case for a number of the volumes in the series: Many works have remained in print on an on-demand basis, while others have such intrinsic value that the physics community has urged us to extend their life span.

The *Advanced Book Classics* series has been designed to meet this demand. It will keep in print those volumes in *Frontiers in Physics* or its sister series, *Lecture Notes and Supplements in Physics*, that continue to provide a unique account of a topic of lasting interest. And through a sizable printing, these classics will be made available at a comparatively modest cost to the reader.

These lecture notes on Richard Feynman's Caltech course on Quantum Electrodynamics were first published in 1961, as part of the first group of lecture note/reprint volumes to be included in the *Frontiers in Physics* series. As is the case with all of the Feynman lecture note volumes, the presentation in this work reflects his deep physical insight, the freshness and originality of his approach to quantum electrodynamics, and the overall pedagogical wizardry of Richard Feynman. Taken together with the reprints included here of

Feynman's seminal papers on the space-time approach to quantum electrodynamics and the theory of positrons, the lecture notes provide beginning students and experienced researchers alike with an invaluable introduction to quantum electrodynamics and to Feynman's highly original approach to the topic.

David Pines
Urbana, Illinois
December 1997

Preface

The text material herein constitutes notes on the third of a three-semester course in quantum mechanics given at the California Institute of Technology in 1953. Actually, some questions involving the interaction of light and matter were discussed during the preceding semester. These are also included, as the first six lectures. The relativistic theory begins in the seventh lecture.

The aim was to present the main results and calculational procedures of quantum electrodynamics in as simple and straightforward a way as possible. Many of the students working for degrees in experimental physics did not intend to take more advanced graduate courses in theoretical physics. The course was designed with their needs in mind. It was hoped that they would learn how one obtains the various cross sections for photon processes which are so important in the design of high-energy experiments, such as with the synchrotron at Cal Tech. For this reason little attention is given to many aspects of quantum electrodynamics which would be of use for theoretical physicists tackling the more complicated problems of the interaction of pions and nucleons. That is, the relations among the many different formulations of quantum electrodynamics, including operator representations of fields, explicit discussion of properties of the S matrix, etc., are not included. These were available in a more advanced course in quantum field theory. Nevertheless, this course is complete in itself, in much the way that a course dealing with Newton's laws can be a complete discussion of mechanics in a physical sense although topics such as least action or Hamilton's equations are omitted.

The attempt to teach elementary quantum mechanics and quantum electrodynamics together in just one year was an experiment. It was based on the idea that, as new fields of physics are opened up, students must work their way further back, to earlier stages of the educational program. The first two terms were the usual quantum mechanical course using Schiff (McGraw-Hill) as a main reference (omitting Chapters X, XII, XIII, and XIV, relating to quantum electrodynamics). However, in order to ease the transition to the latter part of the course, the theory of propagation and potential scattering was developed in detail in the way outlined in Eqs. 15-3 to 15-5. One other unusual point was made, namely, that the nonrelativistic Pauli equation could be written as on page 6 of the notes.

The experiment was unsuccessful. The total material was too much for one year, and much of the material in these notes is now given after a full year graduate course in quantum mechanics.

The notes were originally taken by A. R. Hibbs. They have been edited and corrected by H. T. Yura and E. R. Huggins.

R. P. FEYNMAN
Pasadena, California
November 1961

Acknowledgments

The publisher wishes to acknowledge the assistance of the American Physical Society in the preparation of this volume, specifically their permission to reprint the three articles from the *Physical Review*.

Contents

Quantum Electrodynamics

Interaction of Light with Matter–Quantum Electrodynamics

First Lecture

The theory of interaction of light with matter is called quantum electrodynamics. The subject is made to appear more difficult than it actually is by the very many equivalent methods by which it may be formulated. One of the simplest is that of Fermi. We shall take another starting point by just postulating for the emission or absorption of photons. In this form it is most immediately applicable.

DISCUSSION OF FERMI'S METHOD†

Suppose all the atoms of the universe are in a box. Classically the box may be treated as having natural modes describable in terms of a distribution of harmonic oscillators with coupling between the oscillators and matter.

The transition to quantum electrodynamics involves merely the assumption that the oscillators are quantum mechanical instead of classical. They then have energies $(n + 1/2)\hbar\omega$, $n = 0, 1 \ldots$, with zero-point energy $1/2\hbar\omega$. The box is considered to be full of photons with a distribution of energies $n\hbar\omega$. The interaction of photons with matter causes the number of photons of type n to increase by ± 1 (emission or absorption).

Waves in a box can be represented as plane standing waves, spherical waves, or plane running waves $\exp(i\mathbf{K}\cdot\mathbf{x})$. One can say there is an *instan-*

† Revs. Modern Phys., **4**, 87 (1932).

taneous Coulomb interaction e^2/r_{ij} between all charges plus *transverse waves* only. Then the Coulomb forces may be put into the Schrödinger equation directly. Other formal means of expression are Maxwell's equations in Hamiltonian form, field operators, etc.

Fermi's technique leads to an infinite self-energy term e^2/r_{ii}. It is possible to eliminate this term in suitable coordinate systems but then the transverse waves contribute an infinity (interpretation more obscure). This anomaly was one of the central problems of modern quantum electrodynamics.

Second Lecture

LAWS OF QUANTUM ELECTRODYNAMICS

Without justification at this time the "laws of quantum electrodynamics" will be stated as follows:

1. The amplitude that an atomic system will *absorb* a photon during the process of transition from one state to another is *exactly* the same as the amplitude that the same transition will be made under the influence of a potential equal to that of a classical electromagnetic wave representing that photon, provided: (a) the classical wave is normalized to represent an energy density equal to $\hbar\omega$ times the probability per cubic centimeter of finding the photon; (b) the real classical wave is split into two complex waves $e^{-i\omega t}$ and $e^{+i\omega t}$, and only the $e^{-i\omega t}$ part is kept; and (c) the potential acts only *once* in perturbation; that is, only terms to first order in the electromagnetic field strength should be retained.

Replacing the word "absorbed" by "emit" in rule 1 requires only that the wave represented by $\exp(+i\omega t)$ be kept instead of $\exp(-i\omega t)$.

2. The number of states available per cubic centimeter of a given polarization is

$$d^3\mathbf{K}/(2\pi)^3$$

Note this is exactly the same as the number of normal modes per cubic centimeter in classical theory.

3. Photons obey Bose-Einstein statistics. That is, the states of a collection of identical photons must be symmetric (exchange photons, add amplitudes). Also the statistical weight of a state of n identical photons is 1 instead of the classical n!

Thus, in general, a photon may be represented by a solution of the classical Maxwell equations if properly normalized.

Although many forms of expression are possible it is most convenient to describe the electromagnetic field in terms of plane waves. A plane wave can always be represented by a vector potential only (scalar potential made zero by suitable gauge transformation). The vector potential representing a real classical wave is taken as

$$\mathbf{A} = a\mathbf{e} \cos(\omega t - \mathbf{K}\cdot\mathbf{x})$$

We want the normalization of **A** to correspond to unit probability per cubic centimeter of finding the photon. Therefore the average energy density should be $\hbar\omega$.

Now

$$\mathbf{E} = (1/c)(\partial\mathbf{A}/\partial t) = (\omega a/c)\,\mathbf{e} \sin(\omega t - \mathbf{K}\cdot\mathbf{x})$$

and

$$|\mathbf{B}| = |\mathbf{E}|$$

for a plane wave. Therefore the average energy density is equal to

$$(1/8\pi)\overline{(|\mathbf{E}|^2 + |\mathbf{B}|^2)} = (1/4\pi)(\omega^2 a^2/c^2)\,\overline{\sin^2(\omega t - \mathbf{K}\cdot\mathbf{x})}$$
$$= (1/8\pi)(\omega^2 a^2/c^2)$$

Setting this equal to $\hbar\omega$ we find that

$$a = \sqrt{8\pi\hbar c^2/\omega}$$

Thus

$$\mathbf{A} = \sqrt{8\pi h c^2/\omega}\,\mathbf{e} \cos(\omega t - \mathbf{K}\cdot\mathbf{x})$$
$$= \sqrt{4\pi\hbar c/2\omega}\,\mathbf{e}\,\{\exp[-i(\omega t - \mathbf{K}\cdot\mathbf{x})] + \exp[+i(\omega t - \mathbf{K}\cdot\mathbf{x})]\}$$

Hence we take the amplitude that an atomic system will absorb a photon to be

$$\sqrt{4\pi\hbar c^2/2\omega}\,\exp[-i(\omega t - \mathbf{K}\cdot\mathbf{x})]$$

For emission the vector potential is the same except for a positive exponential.

Example: Suppose an atom is in an excited state Ψ_i with energy E_i and makes a transition to a final state Ψ_f with energy E_f. The probability of transition per second is the same as the probability of transition under the influence of a vector potential $a\mathbf{e}\exp[+i(\omega t - \mathbf{K}\cdot\mathbf{X})]$ representing the emitted photon. According to the laws of quantum mechanics (Fermi's golden rule)

$$\text{Trans. prob./sec} = 2\pi/\hbar\,|_f(\text{potential})_i|^2 \cdot (\text{density of states})$$

$$\text{Density of states} = \frac{K^2\,dK\,d\Omega}{(2\pi)^3\,d(\omega\hbar)} = \frac{\omega^2\,d\Omega}{(2\pi c)^3\,\hbar}$$

The matrix element $U_{fi} = |_f(\text{potential})_i|^2$ is to be computed from perturbation theory. This is explained in more detail in the next lecture. First, however, we shall note that more than one choice for the potential may give the same physical results. (This is to justify the possibility of always choosing $\phi = 0$ for our photon.)

Third Lecture

The representation of the plane-wave photon by the potentials

$$\mathbf{A}(\mathbf{x}, t) = a\mathbf{e} \exp [-i(\omega t - \mathbf{K} \cdot \mathbf{x})]$$

$$\phi = 0$$

is essentially a choice of "gauge." The fact that a freedom of choice exists results from the invariance of the Pauli equation to the quantum-mechanical gauge transform.

The quantum-mechanical transformation is a simple extension of the classical, where, if

$$\mathbf{E} = -\nabla \phi + \partial\phi/\partial t$$

and

$$\mathbf{B} = \nabla \times \mathbf{A}$$

and if χ is any scalar, then the substitutions

$$\mathbf{A}' = \mathbf{A} + \nabla\chi$$

$$\phi' = \phi + \partial\chi/\partial t$$

leave **E** and **B** invariant.

In quantum mechanics the additional transformation of the wave function

$$\Psi' = e^{-i\chi}\Psi$$

is introduced. The invariance of the Pauli equation is shown as follows. The Pauli equation is

$$-\frac{\hbar}{i}\frac{\partial\Psi}{\partial t} = \frac{1}{2m}\left[\sigma \cdot \left(p - \frac{e}{c}A\right)\right]\left[\sigma \cdot \left(p - \frac{e}{c}A\right)\right]\Psi + e\phi\Psi$$

Then, since

$$\frac{\partial}{\partial x}\Psi' = \frac{\partial}{\partial x} e^{-i\chi}\Psi = e^{-i\chi}\frac{\partial \Psi}{\partial x} - i\frac{\partial \chi}{\partial x}\Psi e^{-i\chi}$$

$$p(e^{-i\chi}\Psi) = e^{-i\chi}(p - \nabla\chi)\Psi$$

and

$$\left(p - \frac{e}{c}A\right)e^{-i\chi}\Psi = e^{-i\chi}\left(p - \nabla\chi - \frac{e}{c}A\right)\Psi$$

The partial derivative with respect to time introduces a term $(\partial\chi/\partial t)\Psi e^{-i\chi}$, and this may be included with $\phi e^{-i\chi}\Psi$. Therefore the substitutions

$$\Psi' = e^{-i\chi}\Psi$$

$$A' = A + \frac{e}{c}\nabla\chi$$

$$\phi' = \phi + (\partial\chi/\partial t)$$

leave the Pauli equation unchanged.

The vector potential **A** as defined for a photon enters the Pauli Hamiltonian as a perturbation potential for a transition from state i to state f. Any time-dependent perturbation which can be written

$$\Delta H = e^{i\omega t}\, U(x,y,z)$$

results in the matrix element U_{fi} given by

$$U_{fi} = \int \Psi_f^* \Delta H \Psi_i \, d\,\text{vol}$$

$$= \int \phi_f^*(x) \exp[i(E_f/\hbar)t]\, e^{i\omega t}\, U(x) \exp[-i(E_f/\hbar)t]\, \phi_i(x) d\,\text{vol}$$

This expression indicates that the perturbation has the same effect as a time-independent perturbation U(x,y,z) between initial and final states whose energies are, respectively, $E_i - \omega\hbar$ and E_f. As is well known† the most important contribution will come from the states such that $E_f = E_i - \omega\hbar$.

Using the previous results, the probability of a transition per second is

$$P_{fi}\, d\Omega = \frac{2\pi}{\hbar}|U_{fi}|^2 \frac{\omega^2\, d\Omega}{(2\pi)^3}$$

---

†See, for example, L. D. Landau and E. M. Lifshitz, "Quantum Mechanics; Non-Relativistic Theory," Addison-Wesley, Reading, Massachusetts, 1958, Sec. 40.

To determine U_{fi}, write

$$H = \frac{1}{2m}\left(p - \frac{e}{c}A\right)^2 - \frac{e\hbar}{2mc}(\sigma \cdot \nabla \times A) + eV$$

$$= \frac{1}{2m}p \cdot p + eV - \frac{e}{2mc}(p \cdot A + A \cdot p) - \frac{e\hbar}{2mc}(\sigma \cdot \nabla \times A)$$

$$+ \frac{e^2}{2mc^2} A \cdot A$$

Because of the rule that the potential acts only once, which is the same as requiring only first-order terms to enter, the term in $\mathbf{A} \cdot \mathbf{A}$ does not enter this problem. Making use of $\mathbf{A} = \mathbf{a}\mathbf{e} \exp[-i(\omega t - \mathbf{K} \cdot \mathbf{x})]$ and the two operator relations

(1) $$\nabla \times \mathbf{A} = \mathbf{K} \times \mathbf{e}\, e^{+i\mathbf{K}\cdot\mathbf{x}}\, e^{i\omega t}$$

(2) $$p\, e^{+i\mathbf{K}\cdot\mathbf{x}} = e^{+i\mathbf{K}\cdot\mathbf{x}}(p - \hbar\mathbf{K})$$

or

$$p \cdot \mathbf{e}\, e^{+i\mathbf{K}\cdot\mathbf{x}} = e^{+i\mathbf{K}\cdot\mathbf{x}}(p \cdot \mathbf{e} - \hbar\mathbf{K} \cdot \mathbf{e})$$

where $\mathbf{K}\cdot\mathbf{e} = 0$ (which follows from the choice of gauge and the Maxwell equations), we may write

$$U_{fi} = a \int \phi_f{}^*[-(e/2mc)(p \cdot \mathbf{e}\, e^{+i\mathbf{K}\cdot\mathbf{x}} + e^{+i\mathbf{K}\cdot\mathbf{x}}\, \mathbf{e} \cdot p)$$

$$+ (e\hbar i/2mc)\, \sigma \cdot (\mathbf{K} \times \mathbf{e})\, e^{+i\mathbf{K}\cdot\mathbf{x}}]\, \phi_i \, d\,\mathrm{vol}$$

This result is exact. It can be simplified by using the so-called "dipole" approximation. To derive this approximation consider the term $(e/2mc)(p \cdot \mathbf{e}\, e^{+i\mathbf{K}\cdot\mathbf{x}})$, which is the order of the velocity of an electron in the atom, or the current. The exponent can be expanded.

$$e^{+i\mathbf{K}\cdot\mathbf{x}} = 1 + i\mathbf{K}\cdot\mathbf{x} + 1/2(i\mathbf{K}\cdot\mathbf{x})^2 + \cdots$$

$\mathbf{K}\cdot\mathbf{x}$ is of the order a_0/λ, where a_0 = dimension of the atom and λ = wavelength. If $a_0/\lambda \ll 1$, all terms of higher order than the first in a_0/λ may be neglected. To complete the dipole approximation, it is also necessary to neglect the last term. This is easily done since the last term may be taken as the order of $(\hbar K/mc) = (\hbar Kc/mc^2) \approx (mv^2/2mc^2)$. Although such a term is negligible even this is an overestimate. More correctly,

$$(e\hbar i/2mc)\boldsymbol{\sigma}\cdot(\mathbf{K}\times\mathbf{e})\,e^{+i\mathbf{K}\cdot\mathbf{x}} \approx v/c \times [\text{matrix element of } \boldsymbol{\sigma}\cdot(\mathbf{K}\times\mathbf{p})]$$

The matrix element is

$$\int \phi_f{}^*\boldsymbol{\sigma}\cdot(\mathbf{K}\times\mathbf{p})\,\phi_i\,d\,\text{vol}$$

A good approximation allows the separation

$$\phi_f{}^* = \phi_f{}^*(\mathbf{x})U_f{}^*\,(\text{spin})$$

and

$$\phi_i = \phi_i\,(\mathbf{x})U_i{}^*(\text{spin})$$

Then to the accuracy of this approximation the integral is

$$\int \phi_f{}^*(\mathbf{x})\,\phi_i\,(\mathbf{x})U_f{}^*(\boldsymbol{\sigma}\cdot(\mathbf{K}\times\mathbf{p}))U_i\,d\,\text{vol} = 0$$

since the states are orthogonal.

For the present, the dipole approximation is to be used. Then

$$U_{fi} = -a\frac{e}{c}\frac{\mathbf{p}_{fi}\cdot\mathbf{e}}{m}$$

where

$$\mathbf{p}_{fi}\cdot\mathbf{e} = \int \phi_f{}^*(\mathbf{p}\cdot\mathbf{e})\phi_i = \mathbf{e}\cdot\int \phi_f{}^*\mathbf{p}\phi_i\,d\,\text{vol}$$

So

$$P_{fi} = \frac{2\pi}{\hbar}\left[\frac{e}{mc}\,a\right]^2(\mathbf{p}_{fi}\cdot\mathbf{e})^2\,d\Omega\,\frac{\omega^2}{(2\pi)^3}$$

Using operator algebra, $\mathbf{p}_{fi}/m = \hbar\omega_{fi}\mathbf{x}_{fi}$, so that

$$P_{fi}\,d\Omega = a^2\,[e^2\omega^4/(2\pi)^2](\mathbf{e}\cdot\mathbf{x}_{fi})^2\,d\Omega$$

where $\mathbf{x}_{fi} = \int \phi_f{}^*\mathbf{x}\,\phi_i\,d$ vol. The total probability is obtained by integrating P_{fi} over $d\Omega$, thus

$$\text{Total prob./sec} = \int a^2\,\frac{e^2\omega^4}{(2\pi)^2}(\mathbf{e}\cdot\mathbf{x}_{fi})^2\,d\Omega$$

$$= a^2\,\frac{e^2\omega^4}{2\pi}\int_0^\pi |\mathbf{x}_{fi}|^2\sin^3\theta\,d\theta$$

$$= a^2\,4\,e^2\omega^4\,|\mathbf{x}_{fi}|^2/6\pi$$

The term $\mathbf{e} \cdot \mathbf{x}_{fi}$ is resolved by noting (Fig. 3-1)

$$|\mathbf{x}_{fi} \cdot \mathbf{e}| = |\mathbf{x}_{fi}| \sin\theta$$

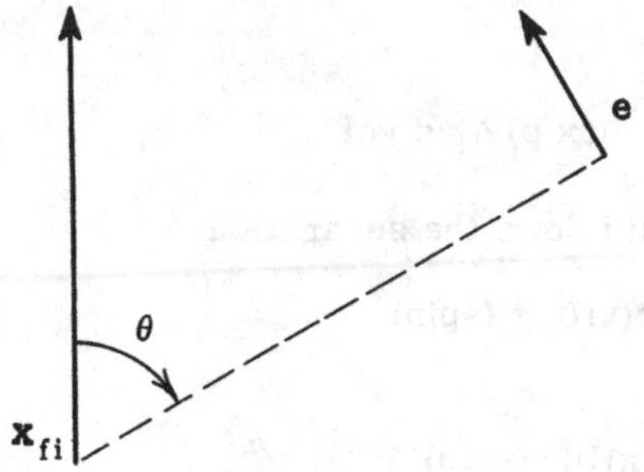

FIG. 3-1

Substituting for a^2,

$$\text{Total prob./sec} = \frac{4}{3}\,\frac{e^2}{\hbar c}\,\frac{\omega^3}{c^2}\,|\mathbf{x}_{fi}|^2$$

Fourth Lecture

Absorption of Light. The amplitude to go from state k to state l in time T (Fig. 4-1) is given from perturbation theory by

$$a_{lk} = -(i/\hbar)\int_0^T \exp\left(\frac{i}{\hbar}E_l t\right) U_{lk}(t) \exp\left(-\frac{i}{\hbar}E_k t\right) dt$$

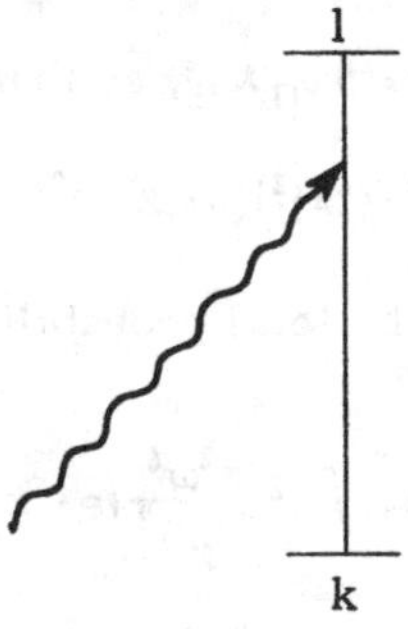

FIG. 4-1

where the time dependence of $U_{kl}(t)$ is indicated by writing

$$U_{lk}(t) = u_{lk}\, e^{-i\omega t}$$

(In accord with the rules of Lecture 2, the argument of the exponential is minus and only terms which are linear in the potential are included.) Using this time dependence and performing the integration,

$$a_{lk} = -\frac{\exp\left[\frac{i}{\hbar}(E_l - \hbar\omega - E_k)\right] - 1}{E_l - \hbar\omega - E_k}\, u_{lk}$$

the transition probability is given by

$$|a_{lk}|^2 = \frac{4\sin^2(\Delta T/2\hbar)}{\Delta^2}|u_{lk}|^2 \qquad \Delta = E_l - E_k - \hbar\omega$$

This is the probability that a photon of frequency ω traveling in direction (θ, ϕ) will be absorbed. The dependence on the photon direction is contained in the matrix element u_{lk}. For example, see Eq. (4-1) for the directional dependence in the dipole approximation.

If the incident radiation contains a range of frequencies and directions, that is, suppose

$$P(\omega,\theta,\phi)d\omega\, d\Omega = \left\{\begin{array}{l}\text{probability that a photon is present with fre-}\\ \text{quency } \omega \text{ to } \omega + d\omega \text{ and in solid angle } d\Omega\\ \text{about the direction } (\theta,\phi)\end{array}\right\}$$

and the probability of absorption of any photon traveling in the (θ,ϕ) direction is desired, it is necessary to integrate over all frequencies. This absorption probability is

$$\int_0^\infty \frac{4\sin^2(\Delta T/2\hbar)}{\Delta^2}|u_{lk}|^2\, P(\omega,\theta,\phi)\, d\omega\, d\Omega$$

when T is large, the factor $(\Delta)^{-2}\sin^2(\Delta T/2\hbar)$ has an appreciable value only for $\hbar\omega$ near $E_l - E_k$, and $P(\omega,\theta,\phi)$ will be substantially constant over the small range in ω which contributes to the integral so that it may be taken out of the integral. Similarly for u_{lk}, so that

$$\text{Trans. prob.} = 2\pi(\hbar)^{-1}|u_{lk}|^2\, P(\omega_{lk},\theta,\phi)d\Omega \qquad (4\text{-}1)$$

where

$$\hbar\omega_{lk} = (E_l - E_k)$$

This can also be written in terms of the incident intensity (energy crossing a unit area in unit time) by noting that

$$\text{Intensity} = i(\omega,\theta,\phi)d\omega\, d\Omega = \hbar\omega c\, P(\omega,\theta,\phi)\, d\omega\, d\Omega$$

Thus

$$\text{Trans. prob.} = 2\pi(\hbar)^{-1}\,|u_{lk}|^2\,(\hbar\omega_{lk}c)^{-1}\, i(\omega_{lk},\theta,\phi)d\Omega \qquad (4\text{-}2)$$

Using the dipole approximation, in which

$$u_{lk} = \sqrt{2\pi/\omega_{lk}}\,(e/mc)(\mathbf{p}_{lk}\cdot\mathbf{e})$$

$$= \sqrt{2\pi/\omega_{lk}}\,(e/c)\,\hbar\omega_{lk}\,(\mathbf{x}_{lk}\cdot\mathbf{e})$$

the total probability of absorption (per second) is

$$4\pi^2 e^2(\hbar c)^{-1}(\mathbf{x}_{lk}\cdot\mathbf{e})^2\, i(\omega_{lk},\theta,\phi)\, d\Omega \qquad (4\text{-}3)$$

It is evident that there is a relation between the probability of spontaneous emission, with accompanying atomic transition from state l to state k,

$$\left\{\begin{matrix}\text{Probability of spontaneous}\\ \text{emission/sec}\end{matrix}\right\} = 2\pi(\hbar)^{-1}(2\pi c)^{-3}\,|u_{kl}|^2\,\omega_{lk}{}^2\, d\Omega$$

and the absorption of a photon with accompanying atomic transition from state k to state l, Eq. (4-1), although the initial and final states are reversed since $|u_{lk}| = |u_{kl}|$. This relation may be stated most simply in terms of the concept of the probability $n(\omega,\theta,\phi)$ that a particular photon state is occupied. Since there are $(2\pi c)^{-3}\omega^2 d\omega\, d\Omega$ photon states in frequency range $d\omega$ and solid angle $d\Omega$, the probability that there is some photon within this range is

$$P(\omega,\theta,\phi)\, d\omega\, d\Omega = n(\omega,\theta,\phi)(2\omega c)^{-3}\omega^2\, d\omega\, d\Omega$$

Expressing the probability of absorption in terms of $n(\omega,\theta,\phi)$,

$$\text{Trans. prob./sec} = 2\pi(\hbar)^{-1}\,|u_{lk}|^2\, n(\omega,\theta,\phi)(2\pi c)^{-3}\,\omega_{kl}{}^2\, d\Omega \qquad (4\text{-}4)$$

This equation may be interpreted as follows. Since $n(\omega,\theta,\phi)$ is the probability that a photon state is occupied, the remainder of the terms of the right-hand side must be the probability per second that a photon in that state will be absorbed. Comparing Eq. (4-4) with the rate of spontaneous emission shows that

$$\left\{\begin{array}{l}\text{Prob./sec of absorption}\\ \text{of a photon from a state}\\ \text{(per photon in that state)}\end{array}\right\} = \left\{\begin{array}{l}\text{prob./sec of spontaneous}\\ \text{emission of a photon into}\\ \text{that state}\end{array}\right\}$$

In what follows, it will be shown that Eq. (4-4) is correct even when there is a possibility of more than one photon per state provided $n(\omega,\theta,\phi)$ is taken as the mean number of photons per state.

If the initial state consists of two photons in the same photon state, it will not be possible to distinguish them and the statistical weight of the initial state will be 1/2! However, the amplitude for absorption will be twice that for one photon. Taking the statistical weight times the square of the amplitude for this process, the transition probability per second is found to be *twice* that for only one photon per photon state. When there are three photons per initial photon state and one is absorbed, the following six processes (shown on Fig. 4-2) can occur.

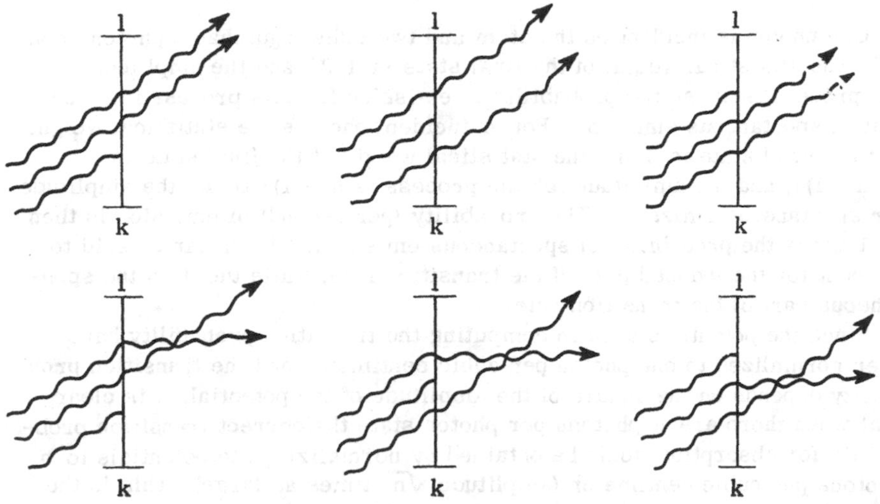

FIG. 4-2

Any of the three incident photons may be absorbed and, in addition, there is the possibility that the photons which are not absorbed may be interchanged. The statistical weight of the initial state is 1/3!, the statistical weight of the final state is 1/2!, and the amplitude for the process is 6. Thus the transition probability is $(1/3!)(1/2!)(6)^2 = 3$ times that if there were one photon per initial state. In general, the transition probability for n photons per initial photon state is n times that for a single photon per photon state, so Eq. (4-4) is correct if $n(\omega,\theta,\phi)$ is taken as the mean number of photons per state.

A transition that results in the emission of a photon may be induced by incident radiation. Such a process (involving one incident photon) could be indicated diagrammatically, as in Fig. 4-3.

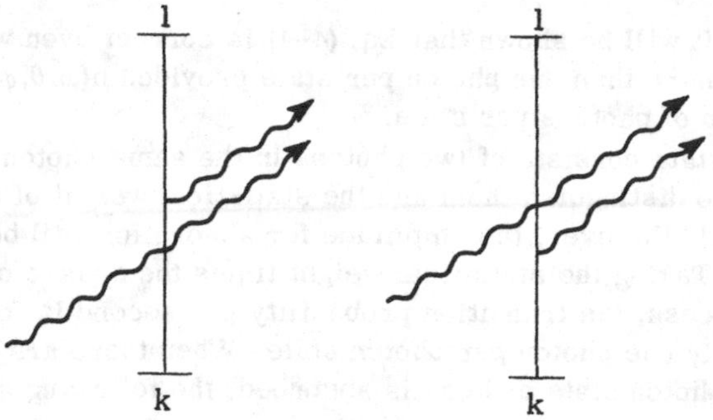

FIG. 4-3

One photon is incident on the atom and two indistinguishable photons come off. The statistical weight of the final state is $1/2!$ and the amplitude for the process is 2, so the probability of emission for this process is twice that of spontaneous emission. For n incident photons the statistical weight of the initial state is $1/n!$, the statistical weight of the final state is $1/(n+1)!$, and the amplitude for the process is $(n+1)!$ times the amplitude for spontaneous emission. The probability (per second) of emission is then $n+1$ times the probability of spontaneous emission. The n can be said to account for the induced part of the transition rate, while the 1 is the spontaneous part of the transition rate.

Since the potentials used in computing the transition probability have been normalized to one photon per cubic centimeter and the transition probability depends on the square of the amplitude of the potential, it is clear that when there are n photons per photon state the correct transition probability for absorption would be obtained by normalizing the potentials to n photons per cubic centimeter (amplitude $\sqrt{n}$ times as large). This is the basis for the validity of the so-called semiclassical theory of radiation. In that theory absorption is calculated as resulting from the perturbation by a potential normalized to the actual energy in the field, that is, to energy $n\hbar\omega$ if there are n photons. The correct transition probability for emission is not obtained this way, however, because it is proportional to $n+1$. The error corresponds to omitting the spontaneous part of the transition probability. In the semiclassical theory of radiation, the spontaneous part of the emission probability is arrived at by general arguments, including the fact that its inclusion leads to the observed Planck distribution formula. Einstein first deduced these relationships by semiclassical reasoning.

Fifth Lecture

Selection Rules in the Dipole Approximation. In the dipole approximation the appropriate matrix element is

$$\mathbf{x}_{if} = \int \Psi_f^* \mathbf{x} \Psi_i \; d\,\text{vol}$$

The components of of $\mathbf{x}_{if}$ are x_{if}, y_{if}, z_{if} and

$$\text{Trans. prob.} \approx |x_{if}|^2 + |y_{if}|^2 + |z_{if}|^2$$

Selection rules are determined by the conditions that cause this matrix element to vanish. For example, if in hydrogen the initial and final states are S states (spherically symmetrical), $\mathbf{X}_{if} = 0$ and transitions between these states are "forbidden." For transitions from P to S states, however, $\mathbf{x}_{if} \neq 0$ and they are "allowed."

In general, for single electron transitions, the selection rule is

$$\Delta L = \pm 1$$

This may be seen from the fact that the coordinates x, y, and z are essentially the Legendre polynomial P_1. If the orbital angular momentum of the initial state is n, the wave function contains P_n. But

$$P_1 P_n = [1/(2n+1)][n P_{n-1} + (n+1)P_{n+1}]$$

Hence for the matrix element not to vanish, the angular momentum of the final state must be $n \pm 1$, so that its wave function will contain either P_{n+1} or P_{n-1}.

For a complex atom (more than one electron), the Hamiltonian is

$$H = \sum_\alpha (1/2m)[\mathbf{P}_\alpha - (e/c)A(\mathbf{x}_\alpha)]^2 + \text{Coulomb terms}$$

The transition probability is proportional to $|P_{mn}|^2 = |\sum_\alpha (P_\alpha)_{mn}|^2$, where the sum is over all the electrons of the atom. As has been shown, $(\mathbf{P}_\alpha)_{mn}$ is the same, up to a constant, as $(\mathbf{x}_\alpha)_{mn}$, and the transition probability is proportional to

$$|\mathbf{x}_{mn}|^2 = \left|\sum_\alpha (\mathbf{x}_\alpha)_{mn}\right|^2$$

In particular, for two electrons the matrix element is

$$\int \Psi_f^*(\mathbf{x}_1,\mathbf{x}_2)(\mathbf{x}_1 + \mathbf{x}_2)\Psi_i(\mathbf{x}_1,\mathbf{x}_2)\, d\mathbf{x}_1\, d\mathbf{x}_2$$

$\mathbf{x}_1 + \mathbf{x}_2$ behaves under rotation of coordinates similarly to the wave function of some "object" with unit angular momentum. If the "object" and the atom in the initial state do not interact, then the product $(\mathbf{x}_1 + \mathbf{x}_2)\Psi_i(\mathbf{x}_1,\mathbf{x}_2)$ can be formally regarded as the wave function of a system (atom + object) having possible values of J_i+1, J_i, and J_i-1 for total angular momentum. Therefore the matrix element is nonzero only if J_f, the final angular momentum, has one of the three values $J_i \pm 1$ or J_i. Hence the general selection rule $\Delta J = \pm 1, 0$.

Parity. Parity is the property of a wave function referring to its behavior upon reflection of all coordinates. That is, if

$$\Psi(-\mathbf{x}_1, -\mathbf{x}_2, \ldots) = +\Psi(\mathbf{x}_1,\mathbf{x}_2, \ldots)$$

parity is even; or if

$$\Psi(-\mathbf{x}_1, -\mathbf{x}_2, \ldots) = -\Psi(\mathbf{x}_1,\mathbf{x}_2, \ldots)$$

parity is odd.

If in the matrix elements involved in the dipole approximation one makes the change of variable of integration $\mathbf{x} = -\mathbf{x}'$, the result is

$$\mathbf{x}_{if} = \int \Psi_f^*(\mathbf{x})\,\mathbf{x}\,\Psi_i(\mathbf{x})\,d^3\mathbf{x} = \int \Psi_f^*(-\mathbf{x}')(-\mathbf{x}')\Psi_0(-\mathbf{x}')\,d^3\mathbf{x}'$$

If the parity of Ψ_f is the same as that of Ψ_i, it follows that

$$\mathbf{x}_{if} = -\mathbf{x}_{if} = 0$$

Hence the rule that *parity must change* in allowed transitions. For a one-electron atom, L determines the parity; therefore, $\Delta L = 0$ would be forbidden. In many-electron atoms, L does not determine the parity (determined by algebraic, not vector, sum of individual electron angular momenta), so $\Delta L = 0$ transitions can occur. The $0 \rightarrow 0$ transitions are always forbidden, however, since a photon always carries one unit of angular momentum.

All wave functions have either even or odd parity. This can be seen from the fact that the Hamiltonian (in the absence of an external magnetic field) is invariant under the parity operation. Then, if $H\Psi(\mathbf{x}) = E\Psi(\mathbf{x})$, it is also true that $H\Psi(-\mathbf{x}) = E\Psi(-\mathbf{x})$. Therefore, if the state is nondegenerate, it follows that either $\Psi(-\mathbf{x}) = \Psi(\mathbf{x})$ or $\Psi(-\mathbf{x}) = -\Psi(\mathbf{x})$. If the state is degenerate, it is possible that $\Psi(-\mathbf{x}) \neq \Psi(\mathbf{x})$. But then a complete solution would be one of the linear combinations

$$\Psi(\mathbf{x}) + \Psi(-\mathbf{x}) \qquad \text{even parity}$$

$$\Psi(\mathbf{x}) - \Psi(-\mathbf{x}) \qquad \text{odd parity}$$

Forbidden Lines. Forbidden spectral lines may appear in gases if they are sufficiently rarefied. That is, forbiddenness is not absolute in all cases. It may simply mean that the lifetime of the state is much longer than if it were allowed, but not infinite. Thus, if the collision rate is small enough (collisions of the second kind ordinarily cause de-excitation in forbidden cases), the forbidden transition may have sufficient time to occur.

In the nearly exact matrix element

$$\int \Psi_f^*(\mathbf{e}\cdot\mathbf{p})\, e^{-i\mathbf{K}\cdot\mathbf{x}}\, \Psi_i\, d^3\mathbf{x}$$

the dipole approximation replaces $e^{-i\mathbf{K}\cdot\mathbf{x}}$ by 1. If this vanishes, the transition is forbidden, as described in the foregoing. The next higher or quadrupole approximation would then be to replace $e^{-i\mathbf{K}\cdot\mathbf{x}}$ by $1 - i/\mathbf{K}\cdot\mathbf{x}$, giving the matrix element

$$-i\int \Psi_f^*(\mathbf{e}\cdot\mathbf{p})(\mathbf{K}\cdot\mathbf{x})\Psi_i\, d^3\mathbf{x}$$

For light moving in the z direction and polarized in the x direction, this becomes

$$-iK\int \Psi_f^*(p_x z)\,\Psi_i\, d^3\mathbf{x} = -iK|_f(p_x z)_i|^2$$

and the transition probability is proportional to

$$(K)^2\,|_f(p_x z)_i|^2$$

whereas in the dipole approximation it was proportional to

$$|_f(p_x)_i|^2$$

Therefore the transition probability in the quadrupole approximation is at least of the order of $(Ka)^2 = a^2/\lambda\!\!\!^-$, smaller than in the dipole approximation, where a is of the order of the size of the atom, and $\lambda\!\!\!^-$ the wavelength emitted.

Problem: Show that

$$H(xz) - (xz)H = (\hbar/mi)(p_x z + xp_z)$$

and consequently that

$$[(\hbar/mi)(p_x z + xp_z)]_{mn} = (xz)_{mn}(E_m - E_n)$$

Note that $p_x z$ can be written as the sum

$$p_x z = 1/2(p_x z + xp_z) + 1/2(p_x z - xp_z)$$

From the preceding problem, the first part of $p_x z$ is seen to be equivalent, up to a constant, to xz, which behaves similarly to a wave function for angular momentum 2, even parity. The second part is the angular momentum operator L_y, which behaves like a wave function for angular momentum 1, even parity. Therefore the selection rules corresponding to the first part are seen to be $\Delta J = \pm 2, +1, 0$ with no parity change. This type of radiation is called electric quadrupole. The selection rules for the second part of $p_x z$ are $\Delta J = \pm 1, 0$, no parity change, and the corresponding radiation is called magnetic dipole. Note that unless $\Delta J \pm 2$, the two types of radiation cannot be distinguished by the change in angular momentum or parity. If $\Delta J = \pm 1, 0$, they can only be distinguished by the polarization of the radiation. Both types may occur simultaneously, producing interference.

In the case of electric quadrupole radiation, it is implicit in the rules that $1/2 \rightarrow 1/2$ and $0 \rightarrow 1$ transitions are forbidden (even though ΔJ may be ± 1), since the required change of 2 for the vector angular momentum is impossible in these cases.

Continuing to higher approximations, it is possible by similar reasoning to deduce the vector change in angular momentum, or angular momentum of the photon, and the selection rules for parity change and change of total angular momentum ΔJ associated with the various multipole orders (Table 5-1).

TABLE 5-1. Classification of Transitions and Their Selection Rules

	Multipole	Electric dipole	Magnetic dipole	Electric quadrupole	Magnetic quadrupole	Electric octupole
Character of photon	Angular momentum	1	1	2	2	3
	Parity	Odd	Even	Even	Odd	Odd
Selection rule for emitting system	Parity change	Yes	No	No	Yes	Yes
	Change of total angular momentum ΔJ	$\pm 1, 0$	$\pm 1, 0$	$\pm 2, \pm 1, 0$	$\pm 2, \pm 1, 0$	$\pm 3, \pm 2, \pm 1, 0$
		No $0 \rightarrow 0$	No $0 \rightarrow 0$	No $0 \rightarrow 0$ $\frac{1}{2} \rightarrow \frac{1}{2}$ $0 \rightarrow 1$	No $0 \rightarrow 0$ $\frac{1}{2} \rightarrow \frac{1}{2}$ $0 \rightarrow 1$	No $0 \rightarrow 0$ $\frac{1}{2} \rightarrow \frac{1}{2}$ etc. (see following)

Actually all the implicit selection rules for ΔJ, which become numerous for the higher multipole orders, can be expressed explicitly by writing the selection rule as

$$|J_f - J_i| \leq l \leq J_f + J_i$$

where 2^l is the multipole order or l is the vector change in angular momentum.

It turns out that in so-called parity-favored transitions, wherein the product of the initial and final parities is $(-1)^{J_f - J_i}$ and the lowest possible multipole order is $J_f - J_i$, the transition probabilities for multipole types contained within the dashed vertical lines in Table 5-1 are roughly equal.† In parity-unfavored transitions, where the parity product is $(-1)^{J_f - J_i + 1}$ and the lowest multipole order is $|J_f - J_i| + 1$, this may not be true.

Sixth Lecture

Equilibrium of Radiation. If a system is in equilibrium, the relative number of atoms per cubic centimeter in two states, say l and k, is given by

$$N_l/N_k = e^{-(E_l - E_k)/kT} = e^{-\hbar\omega/kT}$$

according to statistical mechanics, when the energies differ by $\hbar\omega$. Since the system is in equilibrium, the number of atoms going from state k to l per unit time by absorption of photons $\hbar\omega$ must equal the number going from l to k by emission. If n_ω photons of frequency ω are present per cubic centimeter, then probabilities of absorption are proportional to n_ω and probability of emission is proportional to $n_\omega + 1$. Thus

$$N_k n_\omega = N_l (n_\omega + 1)$$

or

$$(n_\omega + 1)/n_\omega = N_k/N_l = e^{\hbar\omega/kT}$$

$$n_\omega = 1/(e^{\hbar\omega/kT} - 1)$$

This is the Planck black-body distribution law.

The Scattering of Light. We discuss here the phenomena of an incident photon being scattered by an atom into a new direction (and possibly energy) (see Fig. 6-1). This may be considered as the absorption of the incoming photon and the emission of a new photon by the atom. The two photons taking part in the phenomenon are represented by the vector potentials.

$$\mathbf{A}_1 = (2\pi/\omega_1)^{1/2}\, \mathbf{e}_1 e^{+i(\omega_1 t - \mathbf{K}\cdot\mathbf{x})}$$

$$\mathbf{A}_2 = (2\pi/\omega_2)^{1/2}\, \mathbf{e}_2 e^{-i(\omega_2 t - \mathbf{K}\cdot\mathbf{x})}$$

The number to be determined is the probability that an atom initially in state k will be left in state l by the action of the perturbation $\mathbf{A} = \mathbf{A}_1 + \mathbf{A}_2$ in the

---

† For nuclei emitting gamma rays this does not seem to be true. For an obscure reason the magnetic radiation predominates for each order of multipole.

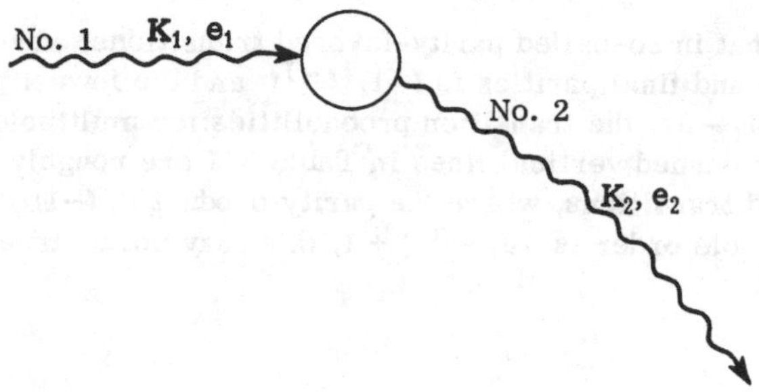

FIG. 6-1

time T. This probability can be computed just as any transition probability with the use of A_{lk}, where

$$A_{lk} = \delta_{kl}\, \exp[-i(E_l/\hbar)T] - (i/\hbar)$$

$$\times \int_0^T \exp[-i(E_l/\hbar)(T - t_3)]\, U_{lk}(t_3)\, \exp[-i(E_k/\hbar)t_3]\, dt_3$$

$$+ \sum_n \int_0^T \int_0^{t_4} \exp[-i(E_l/\hbar)(T - t_4)]$$

$$\times U_{ln}(t_4) \exp[-i(E_n/\hbar)(t_4 - t_3)]\, U_{nk}(t_3)\, \exp[-i(E_k/\hbar)t_3]\, dt_3\, dt_4$$

The dipole approximation is to be employed and

$$U = \Delta H = (e/2mc)(\mathbf{p} \cdot \mathbf{A}) + (e^2/mc^2)(\mathbf{A} \cdot \mathbf{A})$$

where spins are neglected.

In each integral defining A_{lk}, each of the two vector potentials must appear once and only once. Thus, in the first integral the term $\mathbf{p} \cdot \mathbf{A}$ of U will not appear in U_{lk}. The product $\mathbf{A} \cdot \mathbf{A} = (\mathbf{A}_1 + \mathbf{A}_2) \cdot (\mathbf{A}_1 + \mathbf{A}_2)$ will contribute only its cross-product term $2\mathbf{A}_1\mathbf{A}_2$. The second integral will have no contribution from $\mathbf{A} \cdot \mathbf{A}$, but will be the sum of two terms. The first term contains a U_{ln} based on $\mathbf{p} \cdot \mathbf{A}_2$ and a U_{nk} based on $\mathbf{p} \cdot \mathbf{A}_1$. The second has U_{ln} based on $\mathbf{p} \cdot \mathbf{A}_1$ and U_{nk} on $\mathbf{p} \cdot \mathbf{A}_2$. The time sequences resulting in these two terms can be represented schematically as shown in Fig. 6-2.

The integral resulting from the first term will now be developed in detail.

$$(\mathbf{p} \cdot \mathbf{A}_1)_{nk} = (2\pi/\omega_1)^{1/2}\, (\mathbf{p} \cdot \mathbf{e}_1)_{nk}\, e^{-i\omega_1 t}$$

$$(\mathbf{p} \cdot \mathbf{A}_2)_{ln} = (2\pi/\omega_2)^{1/2}\, (\mathbf{p} \cdot \mathbf{e}_2)_{ln}\, e^{i\omega_2 t}$$

Then the resulting integral is

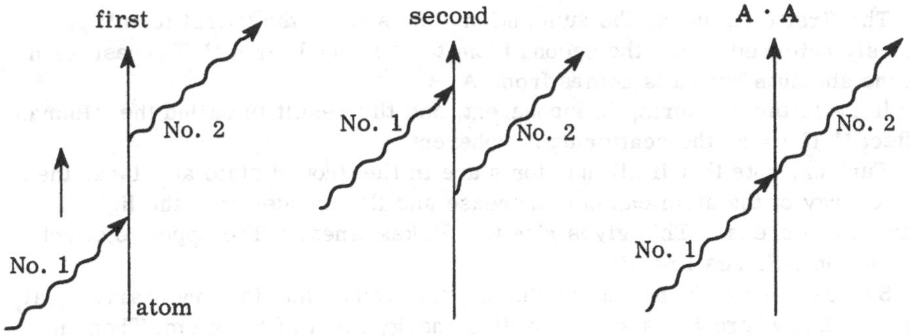

FIG. 6-2

$$\sum_n 2\pi/(\omega_1\omega_2)^{1/2}(p\cdot e_2)_{1n}(p\cdot e_1)_{nk}$$
$$\times \int_0^T\int_0^{t_4} \exp[-i(E_1/\hbar)(T-t_4)+i\omega_2 t_4]$$
$$\times \exp[-i(E_n/\hbar)(t_4-t_3)-i\omega_1 t_3]\exp[-i(E_k/\hbar)t_3]\,dt_3\,dt_4$$

The integral is similar to the integrals considered previously with regard to transition probabilities, and the sum becomes

$$\sum_n 2\pi/(\omega_1\omega_2)^{1/2}(p\cdot e_1)_{1n}(p\cdot e_2)_{nk}\,e^{i\phi}$$
$$\times [\sin(T\cdot\Delta/\hbar)/(E_k-E_n+\hbar\omega_1)\cdot\Delta]$$

where $\Delta=(E_1+\hbar\omega_2-E_k-\hbar\omega_1)$, and the phase angle ϕ is independent of n. A term with the denominator given by $(E_n-\hbar\omega_1-E_k)(E_1+\hbar\omega_2-E_n)$ has been neglected, since previous results show that only energies such that $E_1+\hbar\omega_2\approx E_k+\hbar\omega_1$ are important. The final result can be written

$$\text{Trans. prob./sec} = (2\pi/\hbar)|M|^2[\omega_2^2\,d\Omega_2/(2\pi^3)] \tag{6-1}$$
$$= \sigma c$$

where $|M|$ is determined from A_{1k} by integrating over ω_2 and averaging over e_2. Then the complete expression for the cross section σ is

$$\sigma\,d\Omega_2 = \frac{e^4}{m^2c^4}\frac{\omega_2}{\omega_1}\,d\Omega_2\left|\frac{1}{m}\sum_n \frac{(p\cdot e_2)_{1n}(p\cdot e_1)_{nk}}{E_k+\hbar\omega_1-E_n}\right.$$
$$\left.+\frac{(p\cdot e_1)_{1n}(p\cdot e_2)_{nk}}{E_k-E_n-\hbar\omega_2}+\frac{1}{mc^2}(e_1\cdot e_2)\delta_{1k}\right|^2 \tag{6-2}$$

The first term under the summation comes from the "first term" previously referred to and the second from the "second term." The last term in the absolute brackets comes from $\mathbf{A} \cdot \mathbf{A}$.

If $l \neq k$, the scattering is incoherent, and the result is called the "Raman effect." If $l = k$, the scattering is coherent.

Further, note that if all the atoms are in the ground state and $l \neq k$, then the energy of the atom can only increase and the frequency of the light ω can only decrease. This gives rise to "Stokes lines." The opposite effect gives "anti-Stokes lines."

Suppose $\omega_1 = \omega_2$ (coherent scattering) but further $\hbar\omega_1$ is very nearly equal to $E_k - E_n$, where E_n is some possible energy level of the atom. Then one term in the sum over n becomes extremely large and dominates the remainder. The result is called "resonance scattering." If σ is plotted against ω, then at such values of ω the cross section has a sharp maximum (see Fig. 6-3).

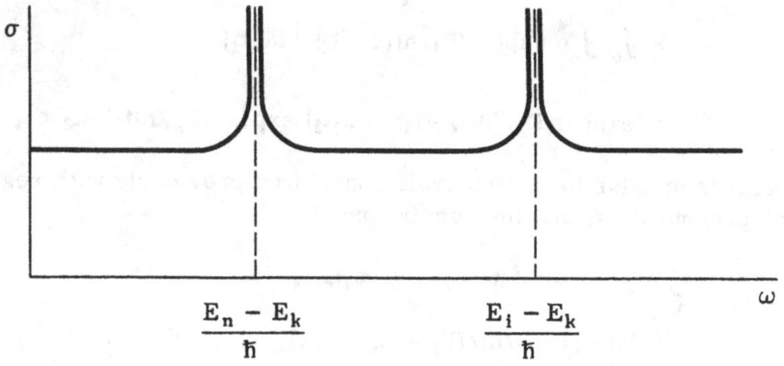

FIG. 6-3

The "index" of refraction of a gas can be obtained by our scattering formula. It can be obtained, as for other types of scattering, by considering the light scattered in the forward direction.

Self-Energy. Another phenomenon that must be considered in quantum electrodynamics is the possibility of an atom emitting a photon and reabsorbing the same photon. This affects the diagonal element A_{kk}. Its effect is equivalent to a shift of energy of the level. One finds

$$\Delta E = \sum_n \int \frac{(\mathbf{p} \cdot \mathbf{e})_{kn} (\mathbf{p} \cdot \mathbf{e})_{nk}}{E_k - E_n - \omega} \frac{d^3\mathbf{K}}{(2\pi\hbar)^3} \frac{2\pi}{\omega}$$

where e is the direction of polarization. This integral diverges. A more exact relativistic calculation also gives a divergent integral. This means that our formulation of electromagnetic effects is not really a completely satisfactory theory. The modifications required to avoid this difficulty of the infinite self-energy will be discussed later. The net result is a very small shift ΔE in position of energy levels. This shift has been observed by Lamb and Rutherford.

Résumé of the Principles and Results of Special Relativity

Seventh Lecture

The principle of relativity is the principle that all physical phenomena would appear to be exactly the same if all the objects concerned were moving uniformly together at velocity v; that is, no experiments made entirely inside of a closed spaceship moving uniformly at velocity v (relative to the center of gravity of the matter in the universe, for example) can determine this velocity. The principle has been verified experimentally. Newton's laws satisfy this principle; for they are unchanged when subject to a Gallilean transformation,

$$x' = x - vt \qquad y' = y \qquad z' = z \qquad t' = t$$

because they involve only second derivatives. The Maxwell equations are changed, however, when subjected to this transformation, and early workers in this field attempted to make an absolute determination of velocity of the earth using this feature (Michelson-Morley experiment). Failure to detect any effects of this type ultimately led to Einstein's postulate that the Maxwell equations are of the same form in any coordinate system; and, in particular, that the velocity of light is the same in all coordinate systems. The transformation between coordinate systems which leaves the Maxwell equations invariant is the Lorentz transformation:

$$x' = \frac{x - vt}{\sqrt{1-(v^2/c^2)}} = x \cosh u - ct \sinh u$$

$$y' = y$$

$$z' = z$$

$$t' = \frac{t - (xv/c^2)}{\sqrt{1 - (v^2/c^2)}} = -\frac{x}{c}\sinh u + t\cosh u$$

where $\tanh u = v/c$. Henceforth we shall use time units so that the speed of light c is unity. The latter form is written to demonstrate the analogy with rotation of axes,

$$x' = x\cos\theta + y\sin\theta$$

$$y' = -x\sin\theta + y\cos\theta$$

Successive transformations v_1 and v_2 or u_1 and u_2 add in the sense that a single transformation v_3 or u_3 will give the same final system if

$$v_3 = v_1 + v_2 \qquad \text{or} \qquad \tanh u_3 = \tanh(u_1 + u_2)$$

Einstein postulated (theory of special relativity) that the Newton laws must be modified in such a way that they, too, are unchanged in form under a Lorentz transformation.

An interesting consequence of the Lorentz transformation is that clocks appear to run slower in moving systems; that is called time dilation. In transforming from one coordinate system to another it is convenient to use tensor analysis. To this end, a four-vector will be defined as a set of four quantities that transforms in the same way as x,y,z and ct. The subscript μ will be used to designate which of the four components is being considered; for example,

$$x_1 = x \qquad x_2 = y \qquad x_3 = z \qquad x_4 = t$$

The following quantities are four-vectors:

$$-\frac{\partial}{\partial x},\ -\frac{\partial}{\partial y},\ -\frac{\partial}{\partial z},\ +\frac{\partial}{\partial t} \qquad (\nabla_\mu) \text{ four-dimensional gradient}$$

$$j_x,\ j_y,\ j_z,\ \rho \qquad (j_\mu) \text{ current (and charge) density}$$

$$A_x,\ A_y,\ A_z,\ \varphi \qquad (A_\mu) \text{ vector (and scalar) potential}$$

$$p_x,\ p_y,\ p_z,\ E \qquad (p_\mu) \text{ momentum and total energy}^\dagger$$

† The energy E, here, is the total energy including the rest energy mc^2.

An invariant is a quantity that does not change under a Lorentz transformation. If a_μ and b_μ are two four-vectors, the "product"

$$a \cdot b \equiv \sum_n a_\mu b_\mu \equiv a_4 b_4 - a_1 b_1 - a_2 b_2 - a_3 b_3$$

is an invariant. To avoid writing the summation symbol, the following summation convention will be used. When the same index occurs twice, sum over it, placing minus in front of first, second, and third components. The Lorentz invariance of the continuity equation is easily demonstrated by writing it as a "product" of four-vectors ∇_μ and j_μ:

$$\nabla_\mu j_\mu = \nabla_4 j_4 - \nabla_1 j_1 - \nabla_2 j_2 - \nabla_3 j_3 = \frac{\partial \rho}{\partial t} + \frac{\partial j_x}{\partial x} + \frac{\partial j_y}{\partial y} + \frac{\partial j_z}{\partial z}$$

Conservation of charge in all systems if it is conserved in one system is a consequence of the invariance of this "product," the four-dimensional divergence $\nabla \cdot j$. Another invariant is

$$p_\mu p_\mu = p \cdot p = E^2 - p_x{}^2 - p_y{}^2 - p_z{}^2 = E^2 - p^2 = m^2$$

(E = total energy, m = rest mass, mc^2 = rest energy, p = momentum.) Thus,

$$E^2 = p^2c^2 + m^2c^4$$

It is also interesting to note that the phase of a free particle wave function $\exp[(-i/\hbar)(Et - \mathbf{p} \cdot \mathbf{X})]$ is invariant since

$$Et - \mathbf{p} \cdot \mathbf{X} = Et - p_x x - p_y y - p_z z = p_\mu p_\mu$$

The invariance of $p_\mu p_\mu$ can be used to facilitate converting laboratory energies to center-of-mass energies (Fig. 6-4) in the following way (consider identical particles, for simplicity):

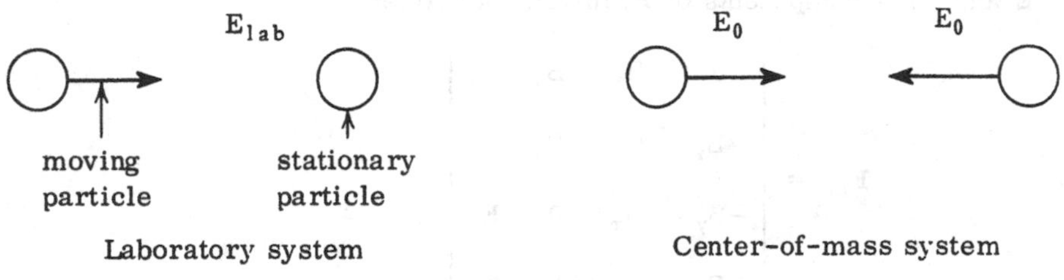

FIG. 6-4

$$p_\mu p_\mu = E_{lab} m = E_0^2 + p_0^2$$

but

$$p_0^2 = E_0^2 - m \qquad \text{so} \qquad E_{lab} m = 2E_0^2 - m^2$$

and

$$E_0 = \left[\frac{1}{2} m (E_{lab} + m)\right]^{1/2}$$

The equations of electrodynamics $\mathbf{B} = \nabla \times \mathbf{A}$ and $\mathbf{E} = -(1/c)(\partial \mathbf{A}/\partial t) - \nabla \phi$ are easily written in tensor notation,

$$B_x = \partial A_z / \partial y - \partial A_y / \partial z = -\nabla_y A_z + \nabla_z A_y$$

$$B_y = \partial A_x / \partial z - \partial A_z / \partial x = -\nabla_z A_x + \nabla_x A_z$$

$$B_z = \partial A_y / \partial x - \partial A_x / \partial y = -\nabla_x A_y + \nabla_y A_x$$

$$E_x = -\partial A_x / \partial t - \partial \phi / \partial x = -\nabla_t A_x + \nabla_x A_t$$

$$E_y = -\partial A_y / \partial t - \partial \phi / \partial y = -\nabla_t A_y + \nabla_y A_t$$

$$E_z = -\partial A_z / \partial t - \partial \phi / \partial z = -\nabla_t A_z + \nabla_z A_t$$

where use is made of the fact that ϕ is the fourth component of the four-vector potential A_μ. From the foregoing it can be seen that B_x, B_y, B_z, E_x, E_y, and E_z are the components of a second-rank tensor:

$$F_{\mu\nu} = \nabla_\mu A_\nu - \nabla_\nu A_\mu \tag{7-1}$$

This tensor is antisymmetric ($F_{\mu\nu} = -F_{\nu\mu}$) and the diagonal terms ($\mu = \nu$) are zero; thus there are only six independent components (three components of **E** and three components of **B**) instead of sixteen.

$$F_{\mu\nu} = \begin{vmatrix} 0 & -B_z & B_y & E_x \\ B_z & 0 & -B_x & E_y \\ -B_y & B_x & 0 & E_z \\ -E_x & -E_y & -E_z & 0 \end{vmatrix}$$

The Maxwell equations $\nabla \times \mathbf{B} = 4\pi \mathbf{J} + (\partial \mathbf{E}/\partial t)$ and $\nabla \cdot \mathbf{E} = 4\pi\rho$ are written

$$\nabla_\mu F_{\mu\nu} = 4\pi j_\nu \tag{7-2}$$

where $\nu = 1, 2, 3, 4$, that is, $j_1 = j_x$, $j_2 = j_y$, $j_3 = j_z$, $j_4 = \rho$, and μ is a dummy index of summation. The $\nu = 1, 2$, and 3 gives the three components of the curl equation, and $\nu = 4$ gives the divergence equation.

The equation satisfied by the potential A_μ is found, by substituting Eq. (7-1) into Eq. (7-2), to be

$$\nabla_\mu \nabla_\mu A_\nu - \nabla_\nu \nabla_\nu A_\mu = 4\pi j_\nu$$

The potential A_ν is not unique, however, since the potential

$$A'_\mu = A_\mu + \nabla_\mu \chi \qquad (7\text{-}3)$$

(χ = any scalar function of position) also satisfies this relation. Such a change or transformation of potential is called a gauge transformation (for historical reasons). We shall make the potentials more definite by assuming that all potentials have been transformed so as to satisfy the so-called Lorentz condition†

$$\nabla_\mu A_\mu = 0 \qquad (7\text{-}4)$$

This is convenient, because it simplifies the equation for A_μ to

$$(\nabla \cdot \nabla) A_\nu = 4\pi j_\nu \qquad (7\text{-}5)$$

since $\nabla \cdot \nabla \equiv \nabla_\mu \nabla_\mu$, which can be recognized as the wave equations

$$\nabla^2 \mathbf{A} - \partial^2 \mathbf{A}/\partial t^2 = -4\pi \mathbf{j} \qquad (7\text{-}5')$$

$$\nabla^2 \phi - \partial^2 \phi/\partial t^2 = -4\pi \rho$$

Sometimes Eq. (7-5′) is written $\Box^2 A_\mu = -4\pi j_\mu$ ($\Box^2$ = D'Alembertian operator $= \nabla^2 - (\partial/\partial t)^2 = -\nabla \cdot \nabla$). This choice of gauge ($\nabla_\mu A_\mu = 0$) is the usual one made in classical electrodynamics,

$$\nabla \cdot \mathbf{A} - \partial\phi/\partial t = 0 \qquad (7\text{-}4')$$

† This is not sufficient to completely define A. We may still use any χ such that $\Box^2 \chi = 0$.

Eighth Lecture

SOLUTION OF THE MAXWELL EQUATION IN EMPTY SPACE

In empty space the plane wave solution of the wave equation

$$\Box^2 A_\mu = -4\pi j_\mu = 0$$

is

$$A_\mu = e_\mu e^{-ik\cdot X}$$

where e_μ and k_μ are constant vectors, and k_μ is subject to the condition that

$$k_\mu k_\mu = k \cdot k = 0$$

This may be seen from the fact that ∇_ν operating on $e^{-ik\cdot X}$ has the effect of multiplying by ik_ν (∇_ν does not operate on e_μ since the coordinates are rectangular). Thus,

$$-\Box^2 A_\mu = \nabla_\nu (\nabla_\nu A_\mu) = \nabla_\nu (-i\, e_\mu k_\nu e^{-ik\cdot X})$$

$$= -e_\mu (k_\nu k_\nu) e^{-ik\cdot X}$$

Note that in these operations $\nabla_\nu A_\mu$ actually forms a second-rank tensor, $\nabla_\nu (\nabla_\nu A_\mu)$ a third-rank tensor, and then contraction on the index ν yields a first-rank tensor or vector.

The k_μ is the propagation vector with components

$$k_\mu = \omega,\ K_x,\ K_y,\ K_z = \omega,\ \mathbf{K}$$

so that in ordinary notation

$$\exp(-ik\cdot x) = \exp[-i(\omega t - \mathbf{K}\cdot\mathbf{X})]$$

and the condition $k\cdot k = 0$ means

$$\omega^2 - \mathbf{K}\cdot\mathbf{K} = 0$$

Problem: Show that the Lorentz condition

$$\nabla_\mu A_\mu = 0$$

implies that $k \cdot e = 0$.

When working in three dimensions it is customary to take the polarization vector $\mathbf{e}$ such that $\mathbf{K}\cdot\mathbf{e} = 0$ and to let the scalar potential $\phi = 0$. But this is not a unique condition; that is, it is not relativistically invariant and will be true only in a one-coordinate system. This would seem to be a paradox attaching some uniqueness to the system in which $\mathbf{K}\cdot\mathbf{e} = 0$, a situation incompatible with relativity theory. The "paradox," however, is resolved by the fact that one can always make a so-called gauge transformation, which leaves the field $F_{\mu\nu}$ unaltered but which *does* change $\mathbf{e}$. Therefore, choosing $\mathbf{K}\cdot\mathbf{e} = 0$ in a particular system amounts to selecting the certain gauge.

The gauge transformation, Eq. (7-3), is

$$\mathbf{A}' = \mathbf{A} + \nabla\chi$$

$$\phi' = \phi + (\partial\chi/\partial t)$$

where χ is a scalar. But $\nabla\cdot A = 0$, the Lorentz condition, Eq. (7-4), will still hold if

$$\nabla\cdot A' = \nabla\cdot A + \nabla\bullet\chi = 0$$

or if

$$\Box^2\chi = 0$$

This equation has a solution $\chi = \alpha e^{-ik\cdot X}$, so

$$A'_\mu = A_\mu + \nabla_\mu(\alpha\, e^{-ik\cdot X}) = (e_\mu + \alpha k_\mu)e^{-ik\cdot x}$$

where α is an arbitrary constant. Therefore,

$$e'_\mu = e_\mu + \alpha k_\mu$$

is the new polarization vector obtained by gauge transformation. In ordinary notation

$$\mathbf{e}' = \mathbf{e} + \alpha\mathbf{K}$$

$$e'_4 = e_4 + \alpha\omega$$

Thus, no matter what coordinate system is used,

$$\mathbf{K}\cdot\mathbf{e}' = \mathbf{K}\cdot\mathbf{e} + \alpha\,\mathbf{K}\cdot\mathbf{K} = \mathbf{K}\cdot\mathbf{e} + \alpha\omega^2$$

can be made to vanish by choice of the constant α.

Clearly the field is left unchanged by a gauge transformation for

$$F'_{\mu\nu} = \nabla_\mu A'_\nu - \nabla_\nu A'_\mu = \nabla_\mu A_\nu + \nabla_\mu\nabla_\nu\chi - \nabla_\nu A_\mu - \nabla_\nu\nabla_\mu\chi = F_{\mu\nu}$$

the $\nabla_\mu \nabla_\nu \chi \equiv \nabla_\nu \nabla_\mu \chi$ because the order of differentiations is immaterial.

RELATIVISTIC PARTICLE MECHANICS

The components of ordinary velocity do not transform in such a manner that they can be components of a four-vector. But another quantity

$$dz_\mu/ds = dt/ds,\ dx/ds,\ dy/ds,\ dz/ds$$

where

$$dz_\mu = dt,\ dx,\ dy,\ dz$$

is an element of path of the particle and ds is the proper time defined by

$$ds^2 = dt^2 - dx^2 - dy^2 - dz^2$$

is a four-vector and is called the four-velocity u_μ. Dividing ds^2 by dt^2 gives the relation between proper time and local time to be

$$(ds/dt)^2 = 1 - v^2$$

The components of ordinary velocity are related as follows:

$$dx/ds = (dx/dt)(dt/ds) = v_x/(1 - y^2)^{1/2}$$

$$dy/ds = v_y/(1 - v^2)^{1/2}$$

$$dz/ds = v_z/(1 - v^2)^{1/2}$$

$$dt/ds = 1/(1 - v^2)^{1/2}$$

It is evident that $u_\mu u_\mu = 1$, for

$$u_\mu u_\mu = \frac{1}{1 - v^2} - \frac{v_x{}^2}{1 - v^2} - \frac{v_y{}^2}{1 - v^2} - \frac{v_z{}^2}{1 - v^2} = \frac{1 - v^2}{1 - v^2} = 1$$

The four-momentum is defined

$$p_\mu = mu_\mu = m/(1 - v^2)^{1/2},\ mv_x/(1 - v^2)^{1/2},\ mv_y/(1 - v^2)^{1/2},$$
$$mv_z/(1 - v^2)^{1/2}$$

Note that $p_4 = m/(1 - v^2)^{1/2}$ is the total energy E, so that in ordinary notation the momentum **P** is given by

$$\mathbf{P} = E\mathbf{v}$$

where **v** is the ordinary velocity.

Like the velocity, the components of ordinary force defined by d/dt (momentum) cannot form the components of a four-vector. But the quantity

$$f_\mu = dp_\mu/ds$$

does form a four-vector with the components

$$f_\mu = d/dt\,(mv_\mu/\sqrt{1-v^2})\,dt/ds = F_\mu/\sqrt{1-v^2} \qquad \mu = 1, 2, 3$$

where F_μ is the ordinary force. The fourth component is

$$f_4 = \frac{\text{power}}{\sqrt{1-v^2}} = \frac{\text{rate of change of energy}}{\sqrt{1-v^2}} = \frac{d/dt(m/\sqrt{1-v^2})}{\sqrt{1-v^2}}$$

This is seen from the fact that $m/\sqrt{1-v^2}$ is the total energy and also from the ordinary identity

$$\text{Power} = \mathbf{F}\cdot\mathbf{V} = \left[\frac{d}{dt}\,\frac{m\mathbf{V}}{\sqrt{1-v^2}}\right]\cdot\mathbf{V}$$

$$= \frac{m}{2}\left[\frac{v^2}{(1-v^2)^{3/2}} + \frac{1}{(1-v^2)^{1/2}}\right]\frac{dv^2}{dt}$$

$$= \frac{mv}{(1-v^2)^{3/2}}\,\frac{dv}{dt} = \frac{d}{dt}\,\frac{m}{\sqrt{1-v^2}}$$

Thus the relativistic analogue of the Newton equations is

$$d/ds\,(p_\mu) = f_\mu = m\,d^2z_\mu/ds^2 \tag{8-1}$$

The ordinary Lorentz force is

$$\mathbf{F} = e(\mathbf{E} + \mathbf{v}\times\mathbf{B}) \tag{8-2}$$

and the rate of change of energy is

$$\mathbf{F}\cdot\mathbf{v} = e\mathbf{E}\cdot\mathbf{v}$$

Then from the preceding definition of four-force,

$$\mathbf{f} = e/(1-v^2)^{1/2}(\mathbf{E} + \mathbf{v}\times\mathbf{B})$$

and

$$f_4 = e/(1 - v^2)^{1/2}\, \mathbf{E} \cdot \mathbf{v}$$

Problem: Show that the expressions just given for **f** and f_4 are equivalent to

$$f_\nu = e u_\mu F_{\mu\nu}$$

so that the relativistic analogue of the Newton equation becomes

$$m\, d^2 z_\mu / ds^2 = e(dz_\nu/ds)\, F_{\mu\nu} \tag{8-3}$$

Also show that this implies

$$d/ds\, [(dz_\mu/ds)^2] = 0$$

In ordinary terms the equation of motion is

$$d/dt(m\mathbf{v}/\sqrt{1 - v^2}) = e(\mathbf{E} + \mathbf{v} \times \mathbf{B}) \tag{8-4}$$

It can be shown by direct application of the Lagrange equations

$$d/dt\, (\partial L/\partial v_\mu) - (\partial L/\partial x_\mu) = 0$$

that the Lagrangian

$$L = -m\sqrt{1 - v^2} - e\phi + e\mathbf{A} \cdot \mathbf{v} \tag{8-5}$$

leads to these equations of motion. Also the momenta conjugate to x is given by $\partial L/\partial \mathbf{v}$ or

$$\mathbf{P} = m\mathbf{v}/(1 - v^2)^{1/2} + e\mathbf{A}$$

The corresponding Hamiltonian is

$$H = e\phi + [(\mathbf{P} - e\mathbf{A})^2 + m^2]^{1/2} \tag{8-6}$$

which satisfies $(H - e\phi)^2 - (\mathbf{P} - e\mathbf{A})^2 = m^2$. It is difficult to convert the Hamiltonian idea to a covariant or four-dimensional formulation. But the principle of least action, which states that the action

$$S = \int L\, dt$$

shall be a minimum, will lead to the relativistic form of the equations of motion directly when expressed as

$$S = \int L\,dt = m \int ds + e \int A_\mu\,(dz_\mu/ds)\,ds$$

$$= \int [m(dz/d\alpha \cdot dz/d\alpha)^{1/2} + eA_\mu\ dz_\mu/d\alpha]\,d\alpha$$

Note that by definition

$$(ds/d\alpha)^2 = (dz_\mu/d\alpha)(dz_\mu/d\alpha)$$

It is interesting that another "action," defined

$$S' = m/2 \int (dz_\mu/d\alpha)^2\,d\alpha + e \int A_\mu(z_\mu)(dz_\mu/d\alpha)\,d\alpha$$

leads to the same result as for S in the foregoing.

Problems: (1) Show that the Lagrangian, Eq. (8-5), leads to the equations of motion, Eq. (8-4), and that the corresponding Hamiltonian is Eq. (8-6). Also find the expression for P. (2) Show that $\delta S = 0$ (variation of S), where S is the action just given, leads to the same equations.

Relativistic Wave Equation

Ninth Lecture

UNITS

The following convention will be used hereafter. We define the units of mass and time and length such that

$$c = 1 \qquad (c = 2.99793 \times 10^{10} \text{ cm/sec})$$

$$\hbar = 1 \qquad (\hbar = 1.0544 \times 10^{-27} \text{ erg/sec})$$

Table 9-1 (top of page 35) is given as a useful reference for conversion to customary units.

The following numerical values are useful:

M_p = mass of proton = 1836.1 m = 938.2 Mev
Mass unit of atomic weights = 931.2 Mev
M_H = Mass of hydrogen atom = 1.00815 mass units
M_N = Mass of neutron = 784 kev + M_H
kT = 1 ev when T = 11,606° K
N_a = Avogadro's number = 6.025×10^{23}
$N_a e$ = 96,520 coulombs

KLEIN-GORDON, PAULI, AND DIRAC EQUATIONS

According to relativistic classical mechanics, the Hamiltonian is given by

$$H = \sqrt{(p - eA)^2 + m^2} + e\phi \qquad (9\text{-}1)$$

TABLE 9-1. Notations and Units

Present notation	Meaning	Customary notation	Value
m	Mass of electron	m	
	Energy	mc^2	510.99 kev
	Momentum	mc	1704 gauss cm
	Frequency	$mc^2/\hbar$	
	Wave number	$mc/\hbar$	
1/m	Length (Compton wavelength)/2π	$\hbar/mc$	3.8615×10^{-11}cm
	Time	$\hbar/mc^2$	
e^2	Fine-structure constant (dimensionless)	$e^2/\hbar c$	1/137.038
e^2/m	Classical radius of the electron	e^2/mc^2	2.8176×10^{-11} cm
$1/me^2$	Bohr radius	$a_0 = \hbar^2/me^2$	0.52945 A

If the quantum-mechanical operator $-i\nabla$ is used for **p**, the operation determined by the square root is undefined. Thus the relativistic quantum-mechanical Hamiltonian has not been obtained directly from the classical equation, Eq. (9-1). However, it is possible to define the square of the operator and to write

$$(H - e\phi)^2 - (\mathbf{p} - e\mathbf{A})^2 = m^2$$

Then, if $H = i\partial/\partial t$,

$$[-(\hbar/i)\,\partial/\partial t - e\phi]^2\Psi - [(\hbar/i)(\partial/\partial x) - e/c A_x]^2\Psi - \cdots = m^2\Psi \tag{9-2}$$

where the square of an operator is evaluated by ordinary operator algebra. This equation was first discovered by Schrödinger as a possible relativistic equation. It is usually referred to as the Klein-Gordon equation. In relativistic notation it is

$$(i\nabla_\mu - eA_\mu)(i\nabla_\mu - eA_\mu)\Psi = m^2\Psi \tag{9-2'}$$

This equation does not allow for "spin" and therefore fails to describe the fine structure of the hydrogen spectrum. It is proposed now for application to the π meson, a particle with no spin. To demonstrate its application to the hydrogen atom, let $A = 0$ and $\phi = -Ze/r$, then let $\Psi = \chi(r) \exp(-iEt)$. Then the equation is

$$(E - Ze^2/r)^2\chi + \nabla^2\chi = m^2\chi$$

Let $E = m + W$, where $W \ll m$, and substituting $V = Ze^2/r$,

$$(W - V)\chi + \nabla^2\chi/2m = -(W - V)^2\chi/2m$$

Neglecting the term on the right in comparison with the first term on the left gives the ordinary Schrödinger equation. By using $(W - V)^2/2m$ as a perturbation potential, the student should obtain the fine-structure splitting for hydrogen and compare with the correct values.

Exercise: For the Klein-Gordon equation, let

$$\rho = i(\Psi^*\partial\Psi/\partial t - \Psi\partial\Psi^*/\partial t) - e\phi\Psi\Psi^* = \text{charge density}$$

$$\mathbf{j} = -i(\Psi^*\nabla\Psi - \Psi\nabla\Psi^*) - e\mathbf{A}\Psi\Psi^* = \text{current density}$$

Then show $(\rho, \mathbf{j})$ is a four-vector and show $\nabla_\mu j_\mu = 0$.

The Klein-Gordon equation leads to a result that seemed so unreasonable at the time it was first brought to light that it was considered a valid basis for rejecting the equation. This result is the possibility of negative energy states. To see that the Klein-Gordon equation predicts such energy states, consider the equation for a free particle, which can be written

$$\Box^2\Psi = m^2\Psi$$

where $\Box^2$ is the D'Alembertian operator. In four-vector notation, this equation has the solution $\Psi = A\exp(-ip_\mu x_\mu)$, where $p_\mu p_\mu = m^2$. Then, since

$$p_\mu p_\mu = p_4p_4 - p_xp_x - p_yp_y - p_zp_z = E^2 - \mathbf{p}\cdot\mathbf{p}$$

there results

$$E = \pm(m^2 + \mathbf{p}\cdot\mathbf{p})^{1/2}$$

The apparent impossibility of negative values of E led Dirac to the development of a new relativistic wave equation. The Dirac equation proves to be correct in predicting the energy levels of the hydrogen atom and is the accepted description of the electron. However, contrary to Dirac's original intent, his equation also leads to the existence of negative energy levels, which by now have been satisfactorily interpreted. Those of the Klein-Gordon equation can also be interpreted.

Exercise: Show if $\Psi = \exp(-iEt)\chi(x,y,z)$ is a solution of the Klein-Gordon equation with constant $\mathbf{A}$ and ϕ, then $\Psi = \exp(+iEt)\chi^*$ is a solution with $-\mathbf{A}$ and $-\phi$ replacing $\mathbf{A}$ and ϕ. This indicates one manner in which "negative" energy solutions can be interpreted. It is the solution for a particle of opposite charge to the electron, but the same mass.

Instead of following the original method in the development of the Dirac equation, a different approach will be used here. The Klein-Gordon equation is actually the four-vector form of the Schrödinger equation. With an analogous point of view, the Dirac equation can be developed as the four-vector form of the Pauli equation.

In following such a procedure, the terms involving "spin" will be included in the relativistic equation. The idea of spin was first introduced by Pauli, but it was not at first clear why the magnetic moment of the electron had to be taken as $\hbar e/2mc$. This value did seem to follow naturally from the Dirac equation, and it is often stated that only the Dirac equation produces as a consequence the correct value of the electron's magnetic moment. However, this is not true, as further work on the Pauli equation showed that the same value follows just as naturally, i.e., as the value that produces the greatest simplification. Because spin is present in the Dirac equation, and absent in the Klein-Gordon, and because the Klein-Gordon equation was thought to be invalid, it is often stated that spin is a relativistic requirement. This is incorrect, since the Klein-Gordon equation is a valid relativistic equation for particles without spin.

Thus the Schrödinger equation is

$$H\Psi = E\Psi$$

where

$$H = 1/2m(-i\nabla - e\mathbf{A})^2 + e\phi$$

and the Klein-Gordon equation is

$$[(H - e\phi)^2 - (-i\nabla - e\mathbf{A})^2]\Psi = m^2\Psi \tag{9-3}$$

Now the Pauli equation is also $H\Psi = E\Psi$, where

$$H = (1/2m)[\boldsymbol{\sigma} \cdot (-i\nabla - e\mathbf{A})]^2 + e\phi \tag{9-4}$$

Thus $(-i\nabla - e\mathbf{A})^2$ appearing in the Schrödinger equation has been replaced by $[\boldsymbol{\sigma} \cdot (-i\nabla - e\mathbf{A})]^2$. Then a possible relativistic version of the Pauli equation, in analogy to the Klein-Gordon equation, might be

$$(H - e\phi)^2\Psi - \{\boldsymbol{\sigma}\cdot[(\hbar/i)\nabla - (e/c)\mathbf{A}]\}^2\Psi = m^2\Psi$$

Actually, this is incorrect, but a very similar form [with H replaced by $i(\partial/\partial t)$] is correct, namely,

$$[i(\partial/\partial t) - e\phi - \boldsymbol{\sigma}\cdot(-i\nabla - e\mathbf{A})] \times [i(\partial/\partial t) - e\phi + \boldsymbol{\sigma}\cdot(-i\nabla - e\mathbf{A})]\,\Psi = m^2\Psi \tag{9-5}$$

This is one form of the Dirac equation.

The wave function Ψ on which the operations are being carried out is actually a matrix:

$$\Psi = \begin{pmatrix}\Psi_+\\ \Psi_-\end{pmatrix}$$

A form closer to that originally proposed by Dirac may be obtained as follows. For convenience, write

$$i(\partial/\partial t) - e\phi = \pi_4$$

$$-i\nabla - (e/c)\mathbf{A} = \boldsymbol{\pi}$$

Now let the function χ be defined by $(\pi_4 + \boldsymbol{\sigma}\cdot\boldsymbol{\pi})\,\Psi = m\chi$.

Then Eq. (9-5) implies $(\pi_4 - \boldsymbol{\sigma}\cdot\boldsymbol{\pi})\chi = m\Psi$. This pair of equations can be rewritten (only to arrive at a particular conventional form) by writing

$$\chi + \Psi = \Psi_a$$

$$\chi - \Psi = \Psi_b$$

Then adding and subtracting the pair of equations for Ψ, χ, there results

$$\pi_4\Psi_a - \boldsymbol{\sigma}\cdot\boldsymbol{\pi}\,\Psi_b = m\Psi_a$$

$$-\pi_4\Psi_b + \boldsymbol{\sigma}\cdot\boldsymbol{\pi}\,\Psi_a = m\Psi_b \tag{9-6}$$

These two equations may be written as one by employing a particular convention. Define a new matrix wave function as

$$\Psi = \begin{pmatrix}\Psi_{a_1}\\ \Psi_{a_2}\\ \Psi_{b_1}\\ \Psi_{b_2}\end{pmatrix} \tag{9-7}$$

where the matrix character of Ψ_a and Ψ_b has been shown explicitly, i.e., actually

$$\Psi_a = \begin{pmatrix} \Psi_{a_1} \\ \Psi_{a_2} \end{pmatrix} \qquad \Psi_b = \begin{pmatrix} \Psi_{b_1} \\ \Psi_{b_2} \end{pmatrix}$$

Then, if the auxiliary definitions are made,

$$\gamma_4 = \left(\begin{array}{cc|cc} 1 & 0 & 0 & 0 \\ 0 & 1 & 0 & 0 \\ \hline 0 & 0 & -1 & 0 \\ 0 & 0 & 0 & -1 \end{array}\right) \qquad \boldsymbol{\gamma} = \left(\begin{array}{cc|cc} 0 & 0 & & \\ 0 & 0 & & \sigma \\ \hline & & 0 & 0 \\ -\sigma & & 0 & 0 \end{array}\right) \tag{9-8}$$

(*Note:* An example of the latter definition is

$$\gamma_x = \begin{pmatrix} 0 & 0 & 0 & 0 \\ 0 & 0 & 1 & 0 \\ 0 & -1 & 0 & 0 \\ -1 & 0 & 0 & 0 \end{pmatrix} \quad \text{since} \quad \sigma_x = \begin{pmatrix} 0 & 1 \\ 1 & 0 \end{pmatrix}$$

γ_y and γ_z are similar.) The two equations in Ψ_a and Ψ_b can be written as one in the form

$$\gamma_4 \pi_4 \Psi - \boldsymbol{\gamma} \cdot \boldsymbol{\pi} \Psi = m\Psi$$

which is actually four equations in four wave functions. Then using four-vector notation, the Dirac equation is

$$\gamma_\mu \pi_\mu \Psi = m\Psi$$

or

$$\gamma_\mu (i\nabla_\mu - eA_\mu)\Psi = m\Psi \tag{9-9}$$

Exercise: Show

$$\gamma_\mu \gamma_\nu + \gamma_\nu \gamma_\mu = \begin{cases} 0 \text{ if } \mu \neq \nu \\ 2 \text{ if } \mu = \nu = 4 \\ -2 \text{ if } \nu = \mu = 1, 2, 3 \end{cases}$$

that is, show

$$\gamma_t^2 = 1 \qquad \gamma_x^2 = \gamma_y^2 = \gamma_z^2 = -1$$

$$\gamma_t \gamma_x = -\gamma_x \gamma_t \qquad \gamma_x \gamma_y = -\gamma_y \gamma_x \qquad \text{etc.}$$

A similar form for the Dirac equation might be obtained by a different argument, by comparison to the Klein-Gordon equation. Thus with $H = i(\partial/\partial t) = i\nabla_4$ and with $e\phi = eA_4$, Eq. (9-3) becomes

$$(i\nabla_\mu - eA_\mu)^2\Psi = m^2\Psi \tag{9-10}$$

in four-vector notation. Using a similar notation in the Pauli equation, Eq. (9-4), but also using $\boldsymbol{\sigma} = \boldsymbol{\gamma}$ and setting $\sigma_4 = \gamma_4$ arbitrarily (to complete the definition of a four-vector form of σ), Eq. (9-4) can be written in a form similar to Eq. (9-10),

$$\{\gamma_\mu[(\hbar/i)\nabla_\mu - (e/c)A_\mu]\}^2\Psi = m^2\Psi \tag{9-11}$$

This should be compared to Eq. (9-9).

Now the Pauli equation, Eq. (9-4), differs from the Schrödinger equation in the replacement of the three-dimensional scalar product $(\mathbf{p} - e\mathbf{A})^2$ by the square of a single quantity $\boldsymbol{\sigma} \cdot (\mathbf{p} - e\mathbf{A})$. Analogously one might guess that the four-vector product $(p_\mu - eA_\mu)^2$ in Eq. (9-10) must be replaced by the square of a single quantity $\gamma_\mu(p_\mu - eA_\mu)$, where we must invent four matrices γ_μ in four dimensions in analogy to the three matrices $\boldsymbol{\sigma}$ in three dimensions. The resulting equation,

$$[\gamma_\mu(i\nabla_\mu - eA_\mu)]^2\Psi = m^2\Psi \tag{9-11}$$

is essentially equivalent to Eq. (9-9) (operate on both sides of Eq. (9-9) by $\gamma_\mu(i\nabla_\mu - eA_\mu)$ and use Eq. (9-9) again to simplify the right-hand side).

Exercise: Show that Eq. (9-11) is equivalent to

$$(i\nabla_\mu - eA_\mu)^2 - \frac{i}{2}e\gamma_\mu\gamma_\nu F_{\mu\nu}\Psi = m^2\Psi$$

Tenth Lecture

ALGEBRA OF THE γ MATRICES

In the preceding lecture the Dirac equation,

$$\gamma_\mu(i\nabla_\mu - eA_\mu)\Psi = m\Psi \tag{10-1}$$

was obtained, together with a special representation for the γ's,

$$\gamma_t = \begin{pmatrix} 1 & 0 \\ 0 & -1 \end{pmatrix} \qquad \gamma_{x,y,z} = \begin{pmatrix} 0 & \sigma_{x,y,z} \\ -\sigma_{x,y,z} & 0 \end{pmatrix} \tag{10-2}$$

where each element in these four-by-four matrices is another two-by-two matrix, that is,

$$1 = \begin{pmatrix} 1 & 0 \\ 0 & 1 \end{pmatrix} \text{ unit matrix} \qquad \sigma_x = \begin{pmatrix} 0 & 1 \\ 1 & 0 \end{pmatrix} \qquad \text{etc.}$$

The best way to define the γ's, however, is to give their commutation relationships, since this is all that is important in their use. The commutation relationships do not determine a unique representation for the γ's, and the foregoing is only one of many possible representations. The commutation relationships are

$$\gamma_t^2 = 1 \qquad \gamma_x^2 = \gamma_y^2 = \gamma_z^2 = -1$$

$$\gamma_t \gamma_{x,y,z} + \gamma_{x,y,z}\gamma_t = 0 \tag{10-3}$$

$$\gamma_x\gamma_y + \gamma_y\gamma_x = 0 \qquad \gamma_x\gamma_z + \gamma_z\gamma_x = 0 \qquad \gamma_y\gamma_z + \gamma_z\gamma_y = 0$$

or, in a unified notation,

$$\gamma_\mu\gamma_\nu + \gamma_\nu\gamma_\rho = 2\delta_{\mu\nu} \tag{10-4}$$

$$\begin{aligned} \delta_{\mu\nu} &= 0 & \mu &\neq \nu \\ &= +1 & \mu &= \nu = 4 \\ &= -1 & \mu &= \nu = 1,2,3 \end{aligned}$$

Note that with this definition of $\delta_{\mu\nu}$ and the rule for forming a scalar product,

$$\delta_{\mu\nu}\, a_\nu = a_\mu$$

Other new matrices may arise by forming products of the matrices already defined. For example, the matrices of Eq. (10-5) are products of γ's taken two at a time. The matrices

$$\gamma_x\gamma_y \qquad \gamma_x\gamma_z \qquad \gamma_y\gamma_z \qquad \gamma_x\gamma_t \qquad \gamma_y\gamma_t \qquad \gamma_z\gamma_t$$

are all independent of γ_x, γ_y, γ_z, γ_t. (They cannot be formed by a linear combination of the latter.) Similarly, products of three matrices,

$$\begin{aligned} &\gamma_x\gamma_y\gamma_z \ (= \gamma_5\gamma_t) \\ &\gamma_y\gamma_z\gamma_t \ (= -\gamma_x\gamma_5) \\ &\gamma_z\gamma_t\gamma_x \ (= -\gamma_y\gamma_5) \\ &\gamma_t\gamma_x\gamma_y \ (= -\gamma_z\gamma_5) \end{aligned}$$

These are the only new products of three. For, if two of the matrices were equal, the product could be reduced, thus $\gamma_t\gamma_y\gamma_t = -\gamma_t\gamma_t\gamma_y = -\gamma_y$. The only new product of four that can be formed is given a special name, γ_5,

$$\gamma_5 \equiv \gamma_x\gamma_y\gamma_z\gamma_t$$

Products of more than four must contain two equal so that they can be reduced. There are, therefore, sixteen linearly independent quantities. Linear combinations of them may involve sixteen arbitrary constants. This agrees with the fact that such a combination can be expressed by a four-by-four matrix. (It is mathematically interesting then that all four-by-four matrices can be expressed in the algebra of the γ's; this is called a Clifford algebra or hypercomplex algebra. A simpler example is that of two-by-two matrices, the so-called algebra of quaternions, which is the algebra of the Pauli spin matrices.)

Exercise: Verify that

$$i\gamma_x\gamma_y = \begin{pmatrix}\sigma_z & 0\\ 0 & \sigma_z\end{pmatrix} \qquad i\gamma_y\gamma_z = \begin{pmatrix}\sigma_x & 0\\ 0 & \sigma_x\end{pmatrix} \qquad i\gamma_z\gamma_x = \begin{pmatrix}\sigma_y & 0\\ 0 & \sigma_y\end{pmatrix} \tag{10-5}$$

and that

$$\gamma_t\gamma_{x,y,z} = \begin{pmatrix}0 & \sigma_{x,y,z}\\ \sigma_{x\ y\ z} & 0\end{pmatrix} \equiv \boldsymbol{\alpha} \text{ (definition of } \boldsymbol{\alpha}\text{)}$$

It is convenient to define another γ matrix, since it occurs frequently:

$$\gamma_5 = \gamma_x\gamma_y\gamma_z\gamma_t = i\begin{pmatrix}0 & 1\\ 1 & 0\end{pmatrix} \tag{10-6}$$

Verify that

$$\gamma_5\gamma_t = i\begin{pmatrix}0 & -1\\ -1 & 0\end{pmatrix} \qquad \gamma_5\gamma_{x,y,z} = -i\begin{pmatrix}\sigma_{x,y,z} & 0\\ 0 & -\sigma_{x,y,z}\end{pmatrix}$$

$$\gamma_5^2 = -1 \qquad \gamma_5\gamma_\mu + \gamma_\mu\gamma_5 = 0 \tag{10-7}$$

For later use, it will be convenient to define

$$\not{a} = a_\mu\gamma_\mu \equiv a_t\gamma_t - a_x\gamma_x - a_y\gamma_y - a_z\gamma_z \tag{10-8}$$

from which it can be shown that

$$\not{a}\not{b} = -\not{b}\not{a} + 2a\cdot b \qquad (a\cdot b = a_\mu b_\mu)$$

$$a^2 = a_\mu a_\mu$$

$$\not{a}\gamma_5 = -\gamma_5\not{a} \tag{10-9}$$

For example, the first may be verified by writing

$$\not{a}\not{b} = (a_t\gamma_t - a_x\gamma_x - a_y\gamma_y - a_z\gamma_z)(b_t\gamma_t - b_x\gamma_x - b_y\gamma_y - b_z\gamma_z)$$

and, moving the second factor to the front, by using the commutation relationships. Doing this with the first term, $(b_t\gamma_t)$ of the second factor produces

$$b_t\gamma_t(a_t\gamma_t + a_x\gamma_x + a_y\gamma_y + a_z\gamma_z)$$

since γ_t commutes with itself and anticommutes with γ_x, γ_y, and γ_z. By performing this operation on all terms, one obtains

$$\begin{aligned}\not{a}\not{b} = b_t\gamma_t[(-a_t\gamma_t + a_x\gamma_x + a_y\gamma_y + a_z\gamma_z) + 2a_t\gamma_t] \\ + b_x\gamma_x[(a_t\gamma_t - a_x\gamma_x - a_y\gamma_y - a_z\gamma_z) + 2a_x\gamma_x] \\ + b_y\gamma_y[(a_t\gamma_t - a_x\gamma_x - a_y\gamma_y - a_z\gamma_z) + 2a_y\gamma_y] \\ + b_z\gamma_z[(a_t\gamma_t - a_x\gamma_x - a_y\gamma_y - a_z\gamma_z) + 2a_z\gamma_z] \\ = -\not{b}\not{a} + 2(b_t a_t\gamma_t^2 + b_x a_x\gamma_x^2 + b_y a_y\gamma_y^2 + b_z a_z\gamma_z^2) \\ = -\not{b}\not{a} + 2b \cdot a\end{aligned}$$

Exercises: (1) Show that

$$\begin{aligned}\gamma_x\not{a}\gamma_x &= \not{a} + 2a_x\gamma_x \\ \gamma_\mu\gamma_\mu &= 4 \\ \gamma_\mu\not{a}\gamma_\mu &= -2\not{a} \\ \gamma_\mu\not{a}\not{b}\gamma_\mu &= 4a \cdot b \\ \gamma_\mu\not{a}\not{b}\not{c}\gamma_\mu &= -2\not{c}\not{b}\not{a}\end{aligned}$$

(2) Verify by expanding in power series that

$$\exp[(u/2)\gamma_t\gamma_x] = \cosh(u/2) + \gamma_t\gamma_x \sinh(u/2)$$

$$\exp[(\theta/2)\gamma_x\gamma_y] = \cos(\theta/2) + \gamma_x\gamma_y \sin(\theta/2) \qquad (10\text{-}10)$$

(3) Show that

$$\exp[-(u/2)\gamma_t\gamma_z]\gamma_t \exp[+(u/2)\gamma_t\gamma_z] = \gamma_t \cosh u + \gamma_z \sinh u$$

$$\exp[-(u/2)\gamma_t\gamma_z]\,\gamma_z\,\exp[+(u/2)\gamma_t\gamma_z] = \gamma_z \cosh u + \gamma_t \sinh u$$

$$\exp[-(u/2)\gamma_t\gamma_z]\,\gamma_y\,\exp[+(u/2)\gamma_t\gamma_z] = \gamma_y$$

$$\exp[-(u/2)\gamma_t\gamma_z]\,\gamma_x\,\exp[+(u/2)\gamma_t\gamma_z] = \gamma_x \qquad (10\text{-}11)$$

EQUIVALENCE TRANSFORMATION

Suppose another representation for the γ's is obtained which satisfies the same commutation relationships, Eq. (10-3); will the form of the Dirac equation, Eq. (10-1), remain the same? To answer this question, make the following transformation of the wave function $\Psi = S\Psi'$, where S is a constant matrix which is assumed to have an inverse S^{-1} ($SS^{-1} = 1$). The Dirac equation becomes

$$\gamma_\mu \pi_\mu S\Psi' = mS\Psi' \qquad (10\text{-}12)$$

The π_μ and S commute, since π is a differential operator plus a function of position, so this equation may be written

$$\gamma_\mu S\pi_\mu \Psi' = mS\,\Psi'$$

Multiplying by the inverse matrix,

$$S^{-1}\gamma_\mu S\pi_\mu \Psi' = mS^{-1}S\Psi'$$

or

$$\gamma'_\mu \pi_\mu \Psi' = m\Psi'$$

where $\gamma'_\mu = S^{-1}\gamma_\mu S$. The transformation $\gamma'_\mu = S^{-1}\gamma_\mu S$ is called an equivalence transformation, and it is easily verified that the new γ's satisfy the commutation relationships, Eq. (10-3). Products of γ's,

$$\gamma'_\mu\gamma'_\nu = (S^{-1}\gamma_\mu S)(S^{-1}\gamma_\nu S) = S^{-1}(\gamma_\mu\gamma_\nu)S$$

transform in exactly the same manner as the γ's, so that equations involving the γ's (the commutation relations specifically) are the same in the transform representation. This demonstrates another representation for the γ's, and the Dirac equation is in exactly the same form as the original, Eq. (10-1), and is equivalent in all its results.

RELATIVISTIC INVARIANCE

The relativistic invariance of the Dirac equation may be demonstrated by assuming, for the moment, that γ transforms similarly to a four-vector.

That is,

$$\gamma_x' = (\gamma_x - v\gamma_t)/(1 - v^2)^{1/2} \qquad \gamma_t' = (\gamma_t - v\gamma_x)/(1 - v^2)^{1/2}$$

$$\gamma_y' = \gamma_y \qquad \gamma_z' = \gamma_z$$

Also π transforms similarly to a four-vector because it is a combination of two four-vectors ∇_μ and A_μ. The left-hand side $\gamma_\mu \pi_\mu$ of the Dirac equation is the product of two four-vectors and hence invariant under Lorentz transformations. The right-hand side m is also invariant. Transforming γ_μ as a four-vector means a new representation for the γ's, but Eqs. (10-11) can be used to show that the new γ's differ from the old γ's by an equivalence transformation; thus it is really not necessary to transform the γ's at all. That is, the same special representation can be used in all Lorentz coordinate systems. This leads to two possibilities in making Lorentz transformations:

1. Transform the γ's similarly to a four-vector and the wave function remains the same (except for Lorentz transformation of coordinates).

2. Use the standard representation in the Lorentz-transformed coordinate system, in which case the wave function will differ from that in (1) by an equivalence transformation.

HAMILTONIAN FORM OF THE DIRAC EQUATION

To show that the Dirac equation reduces to the Schrödinger equation for low velocities, it is convenient to write it in Hamiltonian form. The original term, Eq. (10-1), may be written

$$\gamma_t[-(\hbar/i)(\partial/\partial t) - e\phi]\Psi - \boldsymbol{\gamma} \cdot [(\hbar/i)\boldsymbol{\nabla} - e\mathbf{A}]\Psi = m\Psi$$

Multiplying by $c\gamma_t$ and rearranging terms gives

$$-(\hbar/i)(\partial\Psi/\partial t) = \{\gamma_t \boldsymbol{\gamma} \cdot [(\hbar/i)\boldsymbol{\nabla} - e\mathbf{A}] + e\phi + \gamma_t m\}\Psi$$

$$= H\Psi$$

By Eq. (10-5), H is written

$$H = \boldsymbol{\alpha} \cdot [(\hbar/i)\boldsymbol{\nabla} - e\mathbf{A}] + e\phi + m\beta$$

where $\beta = \gamma_t$, $\alpha_{x,y,z} = \gamma_t\gamma_{x,y,z}$, Eq. (10-5), and the α's satisfy the following commutation relations: $\alpha_x^2 = \alpha_y^2 = \alpha_z^2 = \beta^2 = 1$ and all pairs anticommute.

It will be noted that $\boldsymbol{\alpha}, \beta$ are Hermitian matrices in our special representation, so that in this representation H is Hermitian.

Exercise: Show that a probability density $\rho = \Psi^*\Psi$ and a probability current $\mathbf{j} = \Psi^*\boldsymbol{\alpha}\Psi$ satisfy the continuity equation

$$(\partial\rho/\partial t) + \nabla\cdot\mathbf{j} = 0$$

Note: Ψ is a four-component wave function and

$$\rho = \Psi^*\Psi = (\Psi_1^*\Psi_2^*\Psi_3^*\Psi_4^*)\begin{pmatrix}\Psi_1\\ \Psi_2\\ \Psi_3\\ \Psi_4\end{pmatrix} = \Psi_1^*\Psi_1 + \Psi_2^*\Psi_2 + \Psi_3^*\Psi_3 + \Psi_4^*\Psi$$

$$j_x = \sum_{ij} \Psi_i^*(\alpha_x)_{ij}\,\Psi_j$$

$$= \Psi_4^*\Psi_1 + \Psi_3^*\Psi_2 + \Psi_2^*\Psi_3 + \Psi_1^*\Psi_4$$

Eleventh Lecture

It should be noted that β and α are Hermitian only in certain representations. In particular, they are Hermitian in the representation employed thus far; this will be called the standard representation and expressions in it will be labeled S.R. when appropriate. The Hermitian property of α and β is necessary in order to get

$$\rho = \Psi^*\Psi$$

$$\mathbf{j} = \Psi^*\boldsymbol{\alpha}\Psi \qquad \text{S.R.} \tag{11-1}$$

as the expressions for charge and current density. Hence they are not true in all representations. The Dirac equation is (with $\hbar$, c restored)

$$-(\hbar/i)(\partial\Psi/\partial t) = H\Psi \qquad H = \beta mc^2 + e\phi + c\boldsymbol{\alpha}\cdot[(\hbar/i)\nabla - (e/c)\mathbf{A}] \tag{11-2}†$$

---

† It is noted that the Hamiltonian found in Schiff ("Quantum Mechanics," McGraw-Hill, New York, 1949) differs from this one by negative signs on all but the $e\phi$ term. Also the components Ψ_1, Ψ_2, Ψ_3, Ψ_4 of the wave function used in Schiff correspond, respectively, to $-\Psi_{b_1}$, $-\Psi_{b_2}$, $-\Psi_{a_1}$, Ψ_{a_2} here. All this is the result of an equivalence transformation $S^2 = i\beta\alpha_x\alpha_y\alpha_z$ between the representations used here and in Schiff. It is easily verified that $S^2 = -1$ hence $S^{-1} = -S$ and

$$S^{-1} = \begin{pmatrix} 0 & 0 & 1 & 0\\ 0 & 0 & 0 & 1\\ -1 & 0 & 0 & 0\\ 0 & -1 & 0 & 0\end{pmatrix} = \begin{pmatrix} 0 & 1\\ -1 & 0\end{pmatrix}$$

The expected value of x is

$$\langle x \rangle = \int \Psi^* x \Psi \ d\ vol$$

$$= \int (\Psi_1^* x\Psi_1 + \Psi_2^* x\Psi_2 + \Psi_3^* x\Psi_3 + \Psi_4^* x\Psi_4)\ d\ vol \qquad \text{S.R.}$$

remembering that Ψ now is a four-component wave function. Similarly it may be verified as an exercise that

$$\langle \alpha \rangle = \int \Psi^* \boldsymbol{\alpha} \Psi \ d\ vol$$

$$\langle \alpha_x \rangle = \int (\Psi_4^* \Psi_1 + \Psi_3^* \Psi_2 + \Psi_2^* \Psi_3 + \Psi_1^* \Psi_4)\ d\ vol \qquad \text{S.R.}$$

Also matrix elements are formally the same as before. For example,

$$(\boldsymbol{\alpha})_{mn} = \int \Psi_m^* \boldsymbol{\alpha} \Psi_n \ d\ vol$$

If A is any operator then its time derivative is

$$\dot{A} = i(HA - AH) + \partial A/\partial t$$

For X the result is clearly

$$\dot{\mathbf{x}} = i(H\mathbf{x} - \mathbf{x}H) = \alpha \tag{11-3}$$

since $\mathbf{x}$ commutes with all terms in H except $\mathbf{p}\cdot\alpha$. But $\alpha^2 = 1$, so the eigenvalues of $\boldsymbol{\alpha}$ are ± 1. Hence the eigenvelocities of $\dot{\mathbf{x}}$ are $\pm$ speed of light. This result is sometimes made plausible by the argument that a precise determination of velocity implies precise determinations of position at two times. Then, by the uncertainty principle, the momentum is completely uncertain and all values are equally likely. With the relativistic relation between velocity and momentum, this is seen to imply that velocities near the speed of light are more probable, so that in the limit the expected value of the velocity is the speed of light.†

Similarly,

$$\dot{(\overline{\mathbf{p} - e\mathbf{A}})}_x = i(Hp_x - p_x H) - ie(HA_x - A_x H) - e\partial A_x/\partial t$$

$$= -e(\partial\phi/\partial x) + e\boldsymbol{\alpha}\cdot(\partial \mathbf{A}/\partial x) - e(\boldsymbol{\alpha}\cdot\nabla)A_x - e(\partial A_x/\partial t)$$

The terms in $\mathbf{A}$ and A_x, except the last, expand as follows:

---

† This argument is not completely acceptable, for $\dot{\mathbf{X}}$ commutes with $\mathbf{p}$; that is, one should be able to measure the two quantities simultaneously.

$$e\left(\alpha_x \frac{\partial A_x}{\partial x} + \alpha_y \frac{\partial A_y}{\partial x} + \alpha_z \frac{\partial A_z}{\partial x} - \alpha_x \frac{\partial A_x}{\partial x} - \alpha_y \frac{\partial A_y}{\partial y} - \alpha_z \frac{\partial A_x}{\partial z}\right)$$

This seen to be the x component of

$$e\boldsymbol{\alpha}\times(\nabla \times \mathbf{A}) = e\boldsymbol{\alpha} \times \mathbf{B}$$

The first and last terms form the x component of **E**. Therefore,

$$\overset{\bullet}{\overline{(\mathbf{p} - e\mathbf{A})}} = e(\mathbf{E} + \boldsymbol{\alpha} \times \mathbf{B}) = \mathbf{F}$$

where **F** is the analogue of the Lorentz force. This equation is sometimes regarded as the analogue of Newton's equations. But, since there is no direct connection between this equation and $\dot{\mathbf{x}}$, it does not lead directly to Newton's equations in the limit of small velocities and hence is not completely acceptable as a suitable analogue.

The following relations may be verified as true but their meaning is not yet completely understood, if at all:

$$(d/dt)\,[\mathbf{x} + (i/2m)\beta\boldsymbol{\alpha}] = (\beta/m)(\mathbf{p} - e\mathbf{A})$$

$$(d/dt)\,[t + (i/2m)\beta] = (\beta/m)(H - e\phi)$$

$$i(d/dt)(\alpha_x\alpha_y\alpha_z) = -2m\beta\alpha_x\alpha_y\alpha_z$$

$$-(d/dt)(\beta\boldsymbol{\sigma}) = 2(\beta\alpha_x\alpha_y\alpha_z)(\mathbf{p} - e\mathbf{A})$$

where in the last relation $\boldsymbol{\sigma}$ means the matrix

$$\begin{pmatrix} \boldsymbol{\sigma} & 0 \\ 0 & \boldsymbol{\sigma} \end{pmatrix}$$

so that

$$\sigma_z = -i\alpha_x\alpha_y, \text{ etc.}$$

From analogy to classical physics, one might expect that the angular momentum operator is now

$$\mathbf{L} = \mathbf{R} \times (\mathbf{p} - e\mathbf{A})$$

Note that in classical physics

$$\mathbf{p} - e\mathbf{A} = m\mathbf{v}\,(1 - v^2)^{-1/2}$$

From previous results for $\dot{\mathbf{R}}$ and $\overset{\bullet}{\overline{(\mathbf{p} - e\mathbf{A})}}$, the time derivative of **L** may be written

$$\dot{\mathbf{L}} = \dot{\mathbf{R}} \times (\mathbf{p} - e\mathbf{A}) + \mathbf{R} \times \overline{(\mathbf{p} - e\mathbf{A})}^{\cdot}$$

$$= \boldsymbol{\alpha} \times (\mathbf{p} - e\mathbf{A}) + \mathbf{R} \times \mathbf{F}$$

The last term may be interpreted as torque. For a central force $\mathbf{F}$, this term vanishes. But then it is seen that $\dot{\mathbf{L}} \neq 0$ because of the first term; that is, the angular momentum $\mathbf{L}$ is *not* conserved, even with central forces.

But consider the time derivative of the operator $\boldsymbol{\sigma}$ defined as

$$\begin{pmatrix} \boldsymbol{\sigma} & 0 \\ 0 & \boldsymbol{\sigma} \end{pmatrix}$$

where $\sigma_z = -\alpha_x \alpha_y$, etc. The z component is seen to commute with the β, $e\phi$, and α_z terms of H but not with the α_x and α_y terms, so that $\dot{\sigma}_z = +1(H\alpha_x\alpha_y - \alpha_x\alpha_y H) = +(\alpha_x\pi_x\alpha_x\alpha_y - \alpha_x\alpha_y\alpha_x\pi_x + \alpha_y\pi_y\alpha_x\alpha_y - \alpha_x\alpha_y\alpha_y\pi_y)$, where

$$\boldsymbol{\pi} = (-i\nabla - e\mathbf{A})$$

But

$$\alpha_x\pi_x\alpha_x\alpha_y = \alpha_x\alpha_x\alpha_y\pi_x = \alpha_y\pi_x$$

$$-\alpha_x\alpha_y\alpha_x\pi_x = \alpha_x\alpha_x\alpha_y\pi_x = \alpha_y\pi_x$$

$$\alpha_y\pi_y\alpha_x\alpha_y = -\alpha_y\alpha_y\alpha_x\pi_y = -\alpha_x\pi_y$$

$$-\alpha_x\alpha_y\alpha_y\pi_y = -\alpha_x\pi_y$$

so that

$$\dot{\sigma}_z = (2\alpha_y\pi_x - 2\alpha_x\pi_y)$$

This is seen to be the z component of $-2\boldsymbol{\alpha} \times \boldsymbol{\pi}$. Finally then,

$$1/2(\dot{\boldsymbol{\sigma}}) = -\boldsymbol{\alpha} \times \boldsymbol{\pi} = -\boldsymbol{\alpha} \times (\mathbf{p} - e\mathbf{A})$$

and this is the first term of $\dot{\mathbf{L}}$ with negative sign. Therefore it follows that

$$(d/dt)[\mathbf{L} + (\hbar/2)\boldsymbol{\sigma}] = \mathbf{R} \times \mathbf{F}$$

which vanishes with central forces. The operator $\mathbf{L} + (\hbar/2)\boldsymbol{\sigma}$ may be regarded as the total angular momentum operator, where $\mathbf{L}$ represents orbital angular momentum and $(\hbar/2)\boldsymbol{\sigma}$ intrinsic angular momentum for spin 1/2. Thus total angular momentum is conserved with central forces.

Problems: (1) In a stationary field $\phi = 0$, $\partial A/\partial t = 0$, show that

$$\sigma \cdot (p - eA)$$

is a constant of the motion. Note that this is a consequence of the anomalous gyromagnetic ratio of the electron. It also means that the cyclotron frequency of the electron equals its rate of precession in a magnetic field.

(2) In a stationary magnetic field $\phi = 0$, $\partial A/\partial t = 0$, and for a stationary state, show that Ψ_1, Ψ_2 in

$$\Psi = \begin{pmatrix} \Psi_1 \\ \Psi_2 \\ \Psi_3 \\ \Psi_4 \end{pmatrix}$$

are the same as Ψ_1, Ψ_2 in the Pauli equation. Also, if E_{Pauli} is the kinetic energy in the Pauli equation and $E_{\text{Dirac}} = W + m$ is the rest plus kinetic energy in the Dirac equation, show that

$$E_{\text{Dirac}} = \sqrt{2mE_{\text{Pauli}} + m^2}$$

and explain the simplicity of this relationship.

NONRELATIVISTIC APPROXIMATION TO THE DIRAC EQUATION

It will be assumed that all potentials are stationary and stationary states will be considered. This makes the work simpler but is not necessary. In this case

$$\Psi = e^{-iEt}\,\Psi(X)$$

$$H\Psi = E\Psi \text{ (Dirac Hamiltonian)}$$

and put

$$E = m + W$$

That is,

$$H\Psi = (m + W)\Psi = \alpha \cdot (p - eA)\,\Psi + \beta m\Psi + e\phi\Psi$$

It will be recalled with Ψ written as Eq. (9-5) and with α, β as given in Lecture 10, the previous equation may be written as two equations (9-4′),

$$(m + W)\Psi_a = \sigma \cdot \pi\Psi_b + m\Psi_a + V\Psi_a \tag{11-4}$$

$$(m + W)\Psi_b = \sigma \cdot \pi\Psi_a - m\Psi_b + V\Psi_b \tag{11-5}$$

where, as before, $\pi = (\mathbf{p} - e\mathbf{A})$ and $V = e\phi$. Simplifying and solving Eq. (11-5) for Ψ_b gives

$$\Psi_b = [1/(2m + W - V)]\,(\sigma \cdot \pi)\Psi_a \tag{11-6}$$

It is noted that if W and V are $\ll 2m$, then $\Psi_b \sim (v/c)\Psi_a$. For this reason Ψ_a and Ψ_b are sometimes referred to as the large and small components of Ψ, respectively. Substitution of Ψ_b from Eq. (11-6) into Eq. (11-4) gives

$$W\Psi_a = (\sigma\cdot\pi)[1/(2m + W - V)]\,(\sigma\cdot\pi)\Psi_a + V\Psi_a \tag{11-7}$$

and, if W and V are neglected in comparison to 2m, the result is

$$W\Psi_a = (1/2m)(\sigma\cdot\pi)^2\Psi_a + V\Psi_a$$

This is the Pauli equation, Eq. (9-4).

Now the approximation will be carried out to second order, that is, to order v^2/c^2, to determine just what error may be expected from use of the Pauli equation.

Twelfth Lecture

Using the results of Lecture 11, given by Eqs. (11-6) and (11-7), the low-energy approximation $(w - V) \ll 2m$ will be made, keeping terms to order v^4. Thus

$$(2m + W - V)^{-1} \approx 1/2m - (w - V)/(2m)^2 \tag{12-1}$$

Then Eq. (11-7) becomes

$$(W - V)\Psi_a = (1/2m)(\sigma\cdot\pi)^2\Psi_a - (1/4m^2)(\sigma\cdot\pi)(W - V)(\sigma\cdot\pi)\Psi_a \tag{12-2}$$

while the normalizing requirement $\int(\Psi_a{}^2 + \Psi_b{}^2)\,d\,\text{vol} = 1$, becomes

$$\int \Psi_a{}^*[1 + (\sigma\cdot\pi)^2/(4m^2)]\Psi_a\,d\,\text{vol} = 1 \tag{12-3}$$

By use of the substitution

$$\chi = [1 + (\upsilon \cdot \pi)/(0m^2)]\,\Psi_a \tag{12-4}$$

the normalizing integral can be simplified to read (to order v^2/c^2)

$$\int \chi^* \chi \text{ d vol} = 1$$

This substitution also allows easier interpretation of Eq. (12-2). Rewriting Eq. (12-2),

$$[1 + (\sigma \cdot \pi)^2/(8m^2)]\,(W - V)\,[1 + (\sigma \cdot \pi)^2/(8m^2)]\Psi_a$$

$$= (1/2m)(\sigma \cdot \pi)^2 \Psi_a + (1/8m^2)[(\sigma \cdot \pi)^2(W - V) - 2(\sigma \cdot \pi)(W - V)$$

$$\times\ (\sigma \cdot \pi) + (W - V)(\sigma \cdot \pi)^2]\Psi_a$$

Then applying Eq. (12-4) and dividing by $1 + (\sigma \cdot \pi)^2/(8m^2)$, there results

$$(W - V)\chi = (1/2m)(\sigma \cdot \pi)^2\chi - (1/8m^3)(\sigma \cdot \pi)^4 \chi$$

$$+ (1/8m^2)[(\sigma \cdot \pi)^2 (W - V) - 2(\sigma \cdot \pi)(W - V)(\sigma \cdot \pi)$$

$$+ (W - V)(\sigma \cdot \pi)^2]\,\chi \qquad (12\text{-}5)$$

The techniques of operator algebra may be used to convert Eq. (12-5) to a form more easily interpreted. In particular one should recall that

$$A^2B - 2ABA + BA^2 = A(AB - BA) - (AB - BA)A$$

Then, since $\pi = (\mathbf{p} - e\mathbf{A})$, and since

$$(\sigma \cdot \pi)(W - V) - (W - V)(\sigma \cdot \pi) = -i\,(\sigma \cdot \nabla V)$$

$$= +i\,(\sigma \cdot \mathbf{E})$$

there results [with $\sigma \cdot \pi = A$ and $(W - V) = B$ in the foregoing],

$$i\,(\sigma \cdot \pi)(\sigma \cdot \mathbf{E}) - i\,(\sigma \cdot \mathbf{E})(\sigma \cdot \pi) = \nabla \cdot \mathbf{E} + 2\sigma \cdot (\pi \times \mathbf{E})$$

(since $\nabla \times \mathbf{E} \sim \partial\mathbf{B}/\partial t = 0$ here), so Eq. (12-5) can be expanded as

$$W\chi = \underset{(1)}{V\chi} + \underset{(2)}{(1/2m)(\mathbf{p} - e\mathbf{A})\cdot(\mathbf{p} - e\mathbf{A})\chi} - \underset{(3)}{(e/2m)(\sigma \cdot \mathbf{B})\chi}$$

$$\underset{(4)}{-(1/8m^3)(\mathbf{p} \cdot \mathbf{p})^2 \chi}$$

$$+ \underset{(5)}{(e^2/8m^2)[\nabla \cdot \mathbf{E}} + \underset{(6)}{2\sigma \cdot (\mathbf{p} - e\mathbf{A}) \times \mathbf{E}]\,\chi} \qquad (12\text{-}6)$$

In this form the wave equation may be interpreted by considering each term of Eq. (12-6) separately.

Term (1) gives the ordinary scalar potential energy as it has appeared before.

Term (2) can be interpreted as the kinetic energy.

Term (3), the Pauli spin effect, is just as it appears in the Pauli equation.

Term (4) is a relativistic correction to the kinetic energy. The correction derives from

$$E = (m^2 + p^2)^{1/2} = m(1 + p^2/m^2)^{1/2}$$

$$= m + p^2/2m - p^4/8m^3 + \cdots$$

The last term in this expansion is equivalent to term (4).

Terms (5) and (6) express the spin-orbit coupling. To understand this interpretation consider the part of term (6) given by $\boldsymbol{\sigma} \cdot (\mathbf{p} \times \mathbf{E})$. In an inverse-square field this is proportional to $\boldsymbol{\sigma} \cdot (\mathbf{p} \times \mathbf{r})/r^3$. The factor $\mathbf{p} \times \mathbf{r}$ can be interpreted as the angular momentum $\mathbf{L}$ to get $(\boldsymbol{\sigma} \cdot \mathbf{L})/r^3$, the spin-orbit coupling. This term has no effect when the electron is in a s-state ($L = 0$). On the other hand, (5) reduces to $\boldsymbol{\nabla} \cdot \mathbf{E} = 4\pi Z \delta(r)$, which affects *only* the s-states (when the wave function is nonzero at $r = 0$). So (5) and (6) together result in a continuous function for spin-orbit coupling. The magnetic moment of the electron $e/2m$, appears as the coefficient of term (3), and again of terms (5) and (6), i.e., $(e/2m)(1/4m^2)$.

A classical argument can be made to interpret term (6). A charge moving through an electric field with velocity $\mathbf{v}$ feels an effective magnetic field $\mathbf{B} = \mathbf{v} \times \mathbf{E} = (1/m)(\mathbf{p} - e\mathbf{A}) \times \mathbf{E}$, and term (6) is just the energy $(e/2m) \times (\boldsymbol{\sigma} \cdot \mathbf{B})$ in this field. We get a factor 2 too much this way, however. Even before the development of the Dirac equation, Thomas showed that this simple classical argument is incomplete and gave the correct term (6). The situation is different for the anomalous moments introduced by Pauli to describe neutrons and protons (see Problem 3 below). In Pauli's modified equation, the anomalous moment does appear with the factor 2 when multiplying terms (5) and (6).

Problems: (1) Apply Eq. (12-6) to the hydrogen atom and correct the energy levels to first order. The results should be compared to the exact results.† Note the difference of the wave functions at the origin of coordinates. This difference actually is too restricted in space to have any importance. Near the origin the correct solution to the Dirac equation is proportional to

$$r\,[1 - (Z/137)^2]^{1/2} \approx r^{-1/40{,}000}$$

for the hydrogenic atoms, while the Schrödinger equation gives $\Psi \rightarrow$ constant as $r \rightarrow 0$.

---

†Schiff, "Quantum Mechanics," McGraw-Hill, New York, 1949, pp. 323ff.

(2) Suppose **A** and ϕ depend on time. Let $W = i\partial/\partial t$ and follow through the procedures of this lecture to the same order of approximation.

(3) Pauli's modified equation can be applied to neutrons and protons. It is obtained by adding a term for anomalous moments to the Dirac equation, thus

$$\gamma_\mu (i\nabla_\mu - eA_\mu)\Psi + \mu\gamma_\mu\gamma_\nu F_{\mu\nu}\Psi = m\Psi$$

Multiplying by β, this may be written in the more familiar "Hamiltonian" expression

$$i(\partial/\partial t)\Psi = H_{\text{Dirac}}\,\Psi + \mu\beta(\sigma\cdot \mathbf{B} - \alpha\cdot \mathbf{E})\Psi$$

Show that the same approximation which led to Eq. (12-6) will now produce the terms

$$[V + 1/2M(p - eA)^2 + (\mu + e/2M)\sigma\cdot\mathbf{B} + (1/8M^3)(p\cdot p)^2 + (1/4M^2)(2\mu + e/2M)(\nabla\cdot\mathbf{E} + 2\sigma\cdot(p - eA)\times\mathbf{E})]\,\Psi \qquad (12\text{-}7)$$

for protons, and a similar expression for neutrons, but with $e = 0$.

(4) Equation (12-7) can be used to interpret electron-neutron scattering in an atom. Most of the scattering of neutrons by atoms is the isotropic scattering from the nucleus. However, the electrons of the atom also scatter, and give rise to a wave which interferes with nuclear scattering. For slow neutrons, this effect is experimentally observed. It is interpreted by term (5) of Eq. (12-6) [as modified in Eq. (12-7) with $e = 0$]. Since the electron charge is present outside the nucleus, $\nabla\cdot\mathbf{E}$ has a value different from 0. Term (5) can be used in a Born approximation to compute the amplitude for neutron-electron scattering. However, when the effect was first discovered, it was explained by the assumption of a neutron-electron interaction given by the potential $c\delta(\mathbf{R})$, where δ is the Dirac δ function and R is the neutron-electron distance.

Compute the scattering amplitude with $c\delta(\mathbf{R})$ by the Born approximation and compare with that given by term (5). Show that

$$c = 4\pi\mu_N e^2/4M_N^2$$

In order to interpret $c\delta(\mathbf{R})$ as a potential, the average potential $\overline{V}$ is defined as that potential which, acting over a sphere of radius e^2/mc^2, would produce the same effect.

Using $\mu_N = -1.9135\, e\hbar/2M_N$, show that the resulting V agrees with experimental results within the stated accuracy, i.e., 4400 ± 400 ev.†

† L. Foldy, Phys. Rev., 87, 693 (1952).

(5) Neglecting terms of order v^2/c^2, show that

$$\int \Psi_f{}^* \alpha f(\mathbf{R}) \Psi_i \, d\,\mathrm{vol}$$

$$\rightarrow \int \chi_f{}^* [(\mathbf{p}f + f\mathbf{p})/2m + (\sigma/2m) \times (\nabla f)] \chi_i \, d\,\mathrm{vol}$$

Solution of the Dirac Equation for a Free Particle

Thirteenth Lecture

It will be convenient to use the form of the Dirac equation with the γ's when solving for the free-particle wave functions

$$\gamma_\mu (i\nabla_\mu - eA_\mu)\Psi = m\Psi$$

Using the definition of Lecture 10, $\not{a} = \gamma_\mu a_\mu$,

$$\not{A} = \gamma_\mu A_\mu = \gamma_t A_t - \gamma_x A_x - \gamma_y A_y - \gamma_z A_z$$

$$\not{\nabla} = \gamma_\mu \nabla_\mu = \gamma_t \nabla_t - \gamma_x \nabla_x - \gamma_y \nabla_y - \gamma_z \nabla_z$$

and the Dirac equation may be written

$$(i\not{\nabla} - e\not{A})\Psi = m\Psi \tag{13-1}$$

(Recall that the quantity $\not{a} = \gamma_\mu a_\mu$ is invariant under a Lorentz transformation.)

It is necessary to put the probability density and current into a four-dimensional form. In the special representation, the probability density and current are given by

$$\rho = \Psi^* \Psi \qquad j = \Psi^* \alpha \Psi$$

If the relativistic adjoint† of Ψ is defined

$$\tilde{\Psi} = \Psi^*\beta \tag{13-2}$$

in the standard representation, then the probability density and current may be written

$$\rho = \tilde{\Psi}\beta\Psi \qquad j_\mu = \tilde{\Psi}\gamma_\mu\Psi$$

To verify this, replace $\tilde{\Psi}$ by $\Psi^*\beta$ and note that $\beta^2 = 1$ and that $\beta\gamma_\mu = \alpha_\mu$.

Exercises: (1) Show that the adjoint of Ψ satisfies

$$\tilde{\Psi}(-i\nabla\!\!\!/ - e A\!\!\!/) = m\tilde{\Psi} \tag{13-3}$$

(2) From Eqs. (13-1) and (13-3) show that $\nabla_\mu j_\mu = 0$ (conservation of probability density).

In general, the adjoint of an operator N is denoted by $\tilde{N}$, and $\tilde{N}$ is the same as N except that the order of all γ's appearing in it is reversed, and each explicit i (not those contained in the γ's) is replaced by $-i$. For example, if $N = \gamma_x\gamma_y$, $\tilde{N} = \gamma_y\gamma_x = -N$. If $N = i\gamma_5 = i\gamma_x\gamma_y\gamma_z\gamma_t$, then $\tilde{N} = -i\gamma_t\gamma_z\gamma_y\gamma_x = -i\gamma_5$. The following property takes the place of the Hermitian property so useful in nonrelativistic quantum mechanics:

$$(\tilde{\Psi}_2 N \Psi_1)^* = (\tilde{\Psi}_1 \tilde{N} \Psi_2) \tag{13-4}$$

For a free particle, there are no potentials, so $A\!\!\!/ = 0$ and the Dirac equation becomes

$$i\nabla\!\!\!/\Psi = m\Psi$$

To solve this, try as a solution

$$\Psi = ue^{-ip\cdot x} = ue^{-ip_\nu x_\nu} \tag{13-6}$$

---

† Ψ is a four-component column vector,

$$\begin{pmatrix} \Psi_1 \\ \Psi_2 \\ \Psi_3 \\ \Psi_4 \end{pmatrix}$$

The adjoint Ψ is the four-component row vector $\widetilde{\Psi_1^*, \Psi_2^* -\Psi_3^* -\Psi_4^*}$ in the standard representation. Multiplication by β changes the sign of the third and fourth components, in addition to changing Ψ^* from a column vector to a row vector.

Ψ is a four-component wave function and what is meant by this trial solution is that *each* of the four components is of this form, that is,

$$\begin{pmatrix} \Psi_1 \\ \Psi_2 \\ \Psi_3 \\ \Psi_4 \end{pmatrix} = \begin{pmatrix} u_1 \\ u_2 \\ u_3 \\ u_4 \end{pmatrix} e^{-ip\cdot x}$$

Thus u_1, u_2, u_3, and u_4 are the components of a column vector, and u is called a Dirac spinor. The problem is now to determine what restrictions must be placed on the u's and p's in order that the trial solution satisfy the Dirac equation. The ∇_μ operation on each component of Ψ multiplies each component by $-ip_\mu$, so that the result of this operation on Ψ produces

$$\nabla_\mu \Psi = \nabla_\mu\, u e^{-ip_\nu x_\nu} = -ip_\mu\, u e^{-ip_\nu x_\nu} = -ip_\mu \Psi$$

so that Eq. (13-5) becomes

$$i\gamma_\mu(-ip_\mu)\Psi = \gamma_\mu p_\mu \Psi = \not{p}\Psi = m\Psi \qquad (13\text{-}7)$$

Thus the assumed solution will be satisfactory if $\not{p}u = mu$. To simplify writing, it will now be assumed that the particle moves in the xy plane, so that

$$p_1 = p_x \qquad p_2 = p_y \qquad p_3 = 0 \qquad p_4 = E$$

Under these conditions, $\not{p} = \gamma_t E - \gamma_y p_y - \gamma_x p_x$. In standard representation

$$\gamma_t = \begin{pmatrix} 1 & 0 & 0 & 0 \\ 0 & 1 & 0 & 0 \\ 0 & 0 & -1 & 0 \\ 0 & 0 & 0 & -1 \end{pmatrix} \qquad \gamma_{x,y} = \begin{pmatrix} 0 & \sigma_{x,y} \\ -\sigma_{x,y} & 0 \end{pmatrix}$$

so $\not{p} - m$ becomes

$$\begin{pmatrix} E-m & 0 & 0 & -p_x+ip_y \\ 0 & E-m & -(p_x+p_y) & 0 \\ 0 & p_x-ip_y & -(E+m) & 0 \\ p_x+ip_y & 0 & 0 & -(E+m) \end{pmatrix} \qquad (13\text{-}8)$$

By components, Eq. (13-7) becomes

$$(E-m)u_1 - (p_x - ip_y)u_4 = 0 \qquad (13\text{-}9a)$$

$$(E-m)u_2 - (p_x + ip_y)u_3 = 0 \qquad (13\text{-}9b)$$

$$(p_x - ip_y)u_2 - (E+m)u_3 = 0 \qquad (13\text{-}9c)$$

$$(p_x + ip_y)u_1 - (E+m)u_4 = 0 \qquad (13\text{-}9d)$$

The ratio u_1/u_4 can be determined from Eq. (13-9a) and also from Eq. (13-9d). These two values must agree in order that Eq. (13-6) be a solution. Thus

$$u_1/u_4 = (p_x - ip_y)/(E - m) = (E + m)/(p_x + ip_y)$$

or

$$p_x^2 + p_y^2 + m^2 = E^2 \tag{13-10}$$

This is not a surprising condition. It states that the p_ν must be chosen so as to satisfy the relativistic equation for total energy.

Similarly, Eqs. (13-9b) and (13-9c) can be solved for u_2/u_3 giving

$$u_2/u_3 = (p_x + ip_y)/(E - m) = (E + m)/(p_x - ip_y)$$

which also leads to condition (13-10).

A more elegant way of obtaining exactly the same condition is to start directly with Eq. (13-7). Then, by multiplying this equation by $\not{p}$ gives

$$\not{p}(\not{p}u) = \not{p}(mu) = m(\not{p}u) = m^2u$$

Using Eq. (10-9),

$$\not{p}\not{p} = p \cdot p = E^2 - p_x^2 - p_y^2$$

so that the condition becomes

$$E^2 - p_x^2 - p_y^2 = m^2 \qquad \text{or} \qquad u = 0$$

The former is the same condition as obtained before, and the latter is a trivial solution (no wave function).

Evidently there are two linearly independent solutions of the free-particle Dirac equation. This is so because substitution of the assumed solution, Eq. (13-6), into the Dirac equation gives only a condition on pairs of the u's, u_1, u_4 and u_2, u_3. It is convenient to choose the independent solutions so that each has two components which are zero. Thus the u's for the two solutions can be taken as

$$\begin{pmatrix} F \\ 0 \\ 0 \\ p_+ \end{pmatrix} \quad \text{and} \quad \begin{pmatrix} 0 \\ F \\ p_- \\ 0 \end{pmatrix} \tag{13-11}$$

where the following notation has been used:

$$F = E + m$$
$$p_+ = p_x + ip_y \qquad (13\text{-}12)$$
$$p_- = p_x - ip_y$$

These solutions are not normalized.

DEFINITION OF THE SPIN OF A MOVING ELECTRON

What do the two linearly independent solutions mean? There must be some physical quantity that can still be specified, which will uniquely determine the wave function. It is known, for example, that in the coordinate system in which the particle is stationary there are two possible spin orientations. Mathematically speaking, the existence of two solutions to the eigenvalue equation $\not{p}u = mu$ implies the existence of an operator that commutes with $\not{p}$. This operator will have to be discovered. Observe that γ_5 anticommutes with $\not{p}$; that is, $\gamma_5\not{p} = -\not{p}\gamma_5$. Also observe that any operator $\not{W}$ will anticommute with $\not{p}$ if $W \cdot p = 0$, because

$$\not{W}\not{p} = -\not{p}\not{W} + 2W \cdot p \qquad (10\text{-}9)$$

The combination $\gamma_5\not{W}$ of these two anticommuting operators is an operator which commutes with $\not{p}$; that is,

$$(\gamma_5\not{W})\not{p} = -\gamma_5\not{p}\not{W} = \not{p}(\gamma_5\not{W})$$

The eigenvalues of the operator $(i\gamma_5\not{W})$ must now be found (the i has been added to make eigenvalues come out real in what follows). Denoting these eigenvalues by s,

$$(i\gamma_5\not{W})u = su \qquad (13\text{-}13)$$

To find the possible values of s, multiply Eq. (13-13) by $i\gamma\not{W}$,

$$(i\gamma_5\not{W})(i\gamma_5\not{W})u = -\gamma_5\not{W}\gamma_5\not{W}u = -W \cdot W u = i\gamma_5 W su = s^2u$$

or

$$-W \cdot W = s^2$$

If $W \cdot W$ is taken to be -1, the eigenvalues of the operator $i\gamma_5\not{W}$ are ± 1. The significance of the choice $W \cdot W = -1$ is as follows: In the system in which the particle is at rest, $p_x = p_y = p_z = 0$ and $p_4 = E$. Then

$$0 = p \cdot W = p_4W_4 \qquad \text{or} \qquad W_4 = 0$$

Thus, $W \cdot W = -\mathbf{W} \cdot \mathbf{W} = -1$ or $\mathbf{W} \cdot \mathbf{W} = 1$. This states that in the coordinate system in which the particle is at rest, W is an ordinary vector (it has zero fourth component) with unit length.

When the particle moves in the xy plane, choose $\not{W}$ to be γ_z, so the operator equation for $i\gamma_5 \not{W}$ becomes

$$i\gamma_5 \gamma_z u = su$$

Using relationships derived in Lecture 10, this becomes, for a stationary particle,†

$$i\gamma_5\gamma_z u = i\gamma_x\gamma_y\gamma_t u = i\gamma_x\gamma_y u = \begin{pmatrix} \sigma_z & 0 \\ 0 & \sigma_z \end{pmatrix} u = su$$

This choice makes $\not{W}$ the σ_z operator, and the relationship with spin is clearly demonstrated. If we define u to satisfy both $\not{p}u = mu$ and $i\gamma_5\not{W}u = su$, this completely specifies u. It represents a particle moving with momentum p_μ and having its spin (in the coordinate system moving with the particle) along the W_μ axis either positive (s = +1) or negative (s = −1).

Exercise: Show that the first of the wave functions, Eq. (13-11), is the s = +1 solution and the second is the s = −1 solution.

Another way of obtaining the wave function for a freely moving electron is to perform an equivalence transformation of the wave function as in Eq. (10-12). If the electron is initially at rest with its spin up or down in the z direction, then the spinor for an electron moving with a velocity v in the spatial direction **k** is

$$u(\mathbf{k}) = Su \qquad u = (2m)^{1/2} u_0 \qquad u_0 = \begin{pmatrix} 1 \\ 0 \\ 0 \\ 0 \end{pmatrix} \text{ or } \begin{pmatrix} 0 \\ 1 \\ 0 \\ 0 \end{pmatrix}$$

[For normalization, see Eq. (13-14).]

From Eq. (10-11), S is given by

$$S = \exp[(u/2)\gamma_t\gamma_k] \qquad \cosh u = 1/(1 - v^2)^{1/2}$$

Now

$$\exp[-(u/2)\gamma_t\gamma_k] = \cosh(u/2) + \gamma_t\gamma_k \sinh(u/2)$$

† For a stationary particle $\gamma_t u = u$.

and

$$(2m)^{1/2} \cosh (u/2) = [m(1 - v^2)^{-1/2} + m]^{1/2} = (E + m)^{1/2}$$

$$(2m)^{1/2} \sinh (u/2) = (E - m)^{1/2}$$

Therefore,

$$u_{(k)} = [(E + m)^{1/2} + \gamma_t \gamma_k (E - m)^{1/2}] u_0$$

Writing $f = (E + m)$, $\alpha = \gamma_t \boldsymbol{\gamma}$, and noting $(E^2 - m^2)^{1/2} = p_k$, we get

$$u_{(k)} = (1/\sqrt{F})(E + m + \alpha \cdot p) u_0$$

For the case that p is in the xy plane, this just gives the result, Eq. (13-11), with a normalization factor $1/\sqrt{F}$.

Noticing that for an electron at rest $\gamma_t u_0 = u_0$, $u_{(k)}$ may be written

$$(1/\sqrt{F})(E\gamma_t - \boldsymbol{\gamma} \cdot \mathbf{P} + m) u_0$$

or

$$u_{(k)} = (1/\sqrt{F})(\not{p} + m) u_0$$

It is clear that this is a solution to the free-particle Dirac equation

$$(\not{p} - m) u_k = 0 \qquad (13\text{-}7)$$

for

$$(\not{p} + m)(\not{p} - m) = p^2 - m^2 = 0 \qquad p^2 = m^2$$

NORMALIZATION OF THE WAVE FUNCTIONS

In nonrelativistic quantum mechanics, a plane wave is normalized to give unity probability of finding the particle in a cubic centimeter, that is, $\Psi^* \Psi = 1$. An analogous normalization for the relativistic plane wave might be something like

$$\Psi^* \Psi = u^* u = \bar{u} \gamma_t u = 1$$

However, $\Psi^* \Psi$ transforms similarly to the fourth component of a four-vector (it is the fourth component of four-vector current), so this normalization would not be invariant. It is possible to make a relativistically invariant normalization by setting $u^* u$ equal to the fourth component of a suitable four-vector. For example, E is the fourth component of the momentum four-vector p_μ, so the wave function could be normalized by

$$\tilde{u}\gamma_t u = 2E$$

The constant of proportionality (2) is chosen for convenience in later formulas. Working out $(\tilde{u}\gamma_t u)$ for the $s = +1$ state,

$$(\tilde{u}\gamma_t u) = \begin{pmatrix} F & 0 & 0 & -p_- \end{pmatrix} \begin{pmatrix} 1 & 0 & 0 & 0 \\ 0 & 1 & 0 & 0 \\ 0 & 0 & -1 & 0 \\ 0 & 0 & 0 & -1 \end{pmatrix} \begin{pmatrix} F \\ 0 \\ 0 \\ p_+ \end{pmatrix} \times C_1^2$$

$$= \begin{pmatrix} F & 0 & 0 & -p_- \end{pmatrix} \begin{pmatrix} F \\ 0 \\ 0 \\ -p_+ \end{pmatrix} \times C_1^2 = (F^2 + p_+p_-)C_1^2 = 2E(E+m)C_1^2$$

The C_1 is the normalizing factor multiplying the wave functions of Eq. (13-11). In order that $(\tilde{u}\gamma_t u)$ be equal to 2E, the normalizing factor must be chosen $(E+m)^{-1/2} = (F)^{-1/2}$. In terms of $(\tilde{u}u)$, this normalizing condition becomes

$$(\tilde{u}u) = \begin{pmatrix} F & 0 & 0 & -p_- \end{pmatrix} \begin{pmatrix} F \\ 0 \\ 0 \\ p_+ \end{pmatrix} \times \frac{1}{F} = (F^2 - p_-p_+)\frac{1}{F}$$

$$= \frac{2m^2 + 2mE}{E+m} = 2m$$

The same result is obtained for the $s = -1$ state. Thus the normalizing condition can be taken as

$$(\tilde{u}u) = 2m \tag{13-14}$$

In a similar manner, the following can be shown to be true:

$$(\tilde{u}\gamma_x u) = 2p_x$$

$$(\tilde{u}\gamma_y u) = 2p_y$$

$$(\tilde{u}\gamma_z u) = 0$$

It will be convenient to have the matrix elements of all the γ's between various initial and final states, so Table 13-1 has been worked out.

TABLE 13-1. Matrix Elements for Particle Moving in the xy Plane

Matrix N	$(\tilde{u}Nu)$ $s = +1$	$\sqrt{F_1F_2}(\tilde{u}_2Nu_1)$ $s_1 = +1$ $s_2 = +1$	$\sqrt{F_1F_2}(\tilde{u}_2Nu_1)$ $s_1 = +1$ $s_2 = -1$	$\sqrt{F_1F_2}(\tilde{u}_2Nu_1)$ $s_1 = -1$ $s_2 = -1$	$\sqrt{F_1F_2}(\tilde{u}_2Nu_1)$ $s_1 = -1$ $s_1 = +1$
1	2m	$F_2F_1 - p_{1+}p_{2-}$	0	Complex conjugate of $s_1 = +1$, $s_2 = +1$ case (column 3)	Negative of complex conjugate of $s_1 = +1$, $s_2 = -1$ case (column 4)
γ_x	$2p_x$	$F_2p_{1+} + p_{2-}F_1$	0		
γ_y	$2p_y$	$-iF_2p_{1+} + ip_{2-}F_1$	0		
γ_z	0	0	$-p_{1+}F_2 + p_{2+}F_1$		
γ_t	2E	$F_2F_1 + p_{1+}p_{2-}$	0		
$\gamma_y\gamma_z$	0	0	$-iF_2F_1 + ip_{1+}p_{2+}$		
$\gamma_z\gamma_x$	0	0	$F_2F_1 + p_{1+}p_{2+}$		
$\gamma_x\gamma_y$	−2iE	$-iF_2F_1 - ip_{1+}p_{2-}$	0		
$\gamma_t\gamma_x$	$2ip_y$	$F_2p_{1+} - p_{2-}F_1$	0		
$\gamma_t\gamma_y$	$-2ip_x$	$-iF_2p_{1+} - ip_{2-}F_1$	0		
$\gamma_t\gamma_z$	0	0	$-p_{1+}F_2 - p_{2+}F_1$		
$\gamma_5\gamma_x = \gamma_t\gamma_y\gamma_z$	0	0	$-iF_2F_1 - ip_{1+}p_{2+}$		
$\gamma_5\gamma_y = \gamma_t\gamma_z\gamma_x$	0	0	$F_2F_1 - p_{1+}p_{2+}$		
$\gamma_5\gamma_z = \gamma_t\gamma_x\gamma_y$	−2im	$-iF_2F_1 + ip_{1+}p_{2-}$	0		
$\gamma_5\gamma_t = \gamma_x\gamma_y\gamma_z$	0	0	$iF_2p_{1+} + iF_1p_{2+}$		
$\gamma_5 = \gamma_x\gamma_y\gamma_z\gamma_t$	0	0	$iF_2p_{1+} - iF_1p_{2+}$		

Note: $p_{2+} = p_{2x} + ip_{2y} = p_2 \exp(i\theta_2)$; $p_{2-} = p_{2x} - ip_{2y} = p_2 \exp(-i\theta_2)$; $F_2 = E_2 + m$; $F_1 = E_1 + m$; $p^2 = (E - m)F$.

Limiting cases: To obtain the case where 1 is a positron at rest, the table gives $\sqrt{F_2}(\bar{u}_2 N u_1)$ if one puts $F_1 = 0$, $p_{1+} = 1 = p_{1-}$ in the table. For both at rest as positrons, the table gives $(\bar{u}_2 N u_1)$ with $F_1 = F_2 = 0$; $p_{1+} = p_{2+} = 1$.

Fourteenth Lecture

METHODS OF OBTAINING MATRIX ELEMENTS

The matrix element of an operator M between initial state u_1 and final state u_2 will be denoted by

$$(\bar{u}_2 M u_1)$$

The matrix element is independent of the representations used if they are related by unitary equivalence transformations. That is,

$$u'_1 = Su_1$$

$$u'_2 = Su_2$$

$$M' = SMS^{-1}$$

$$\bar{u}'_2 = \bar{u}_2\tilde{S}$$

so that

$$\bar{u}'_2 M' u'_1 = \bar{u}_2 \tilde{S} S\, M S^{-1} S u_1 = \bar{u}_2 M u_1$$

where the property $\tilde{S} = S^{-1}$ has been assumed for S.

The straightforward method to compute the matrix elements is simply to write them out in matrix form and carry out the operations. In this way the data in Table 13-1 were obtained.

Other methods may be used, however, sometimes simpler and sometimes leading to corollary information, as illustrated by the following example. By the normalization convention,

$$\bar{u}u = 2m$$

Hence

$$(\bar{u} \not{p} u) = 2m^2$$

since $\not{p}u = mu$. Similarly,

$$(\tilde{u}\gamma_\mu \not{p} u) = m(\tilde{u}\gamma_\mu u)$$

But also note that

$$(\tilde{u}\not{p}\gamma_\mu u) = m(\tilde{u}\gamma_\mu u)$$

because $\tilde{u}\not{p} = \not{p}\tilde{u} = m\tilde{u}$. Adding the two expressions, one obtains

$$(\tilde{u}(\gamma_\mu \not{p} + \not{p}\gamma_\mu)u) = 2m(\tilde{u}\gamma_\mu u)$$

From the relation proved in the exercises that

$$\not{a}\not{b} = -\not{b}\not{a} + 2a \cdot b$$

it is seen that

$$\not{p}\gamma_\mu + \gamma_\mu \not{p} = 2p_\mu \qquad \gamma_\mu = \not{1}$$

But p_μ is just a number, so it follows that

$$2p_\mu (\tilde{u}u) = 2m(\tilde{u}\gamma_\mu u)$$

and since $\tilde{u}u = 2m$, by normalization

$$(\tilde{u}\gamma_\mu u) = 2p_\mu$$

Furthermore, the general relation

$$(\tilde{u}\gamma_t u)/(\tilde{u}u) = p_4/m = E/m$$

is obtained. From this it is seen why the possible normalization

$$(\tilde{u}\gamma_\mu u) = E/m$$

was equivalent to $(\tilde{u}u) = 1$.

Problem: Using methods analogous to the one just demonstrated, show that

$$(\tilde{u}\gamma_5 u) = 0$$

INTERPRETATIONS OF NEGATIVE ENERGY STATES

It was found that a necessary condition for solution of the Dirac equation to exist is

$$E^2 = p^2 + m^2$$

$$E = \pm (p^2 + m^2)^{1/2}$$

The meaning of the positive energy is clear but that of the negative is not. It was at one time suggested by Schrödinger that it should be arbitrarily excluded as having no meaning. But it was found that there are two fundamental objections to the exclusion of negative energy states. The first is physical, theoretically physical, that is. For the Dirac equation yields the result that starting with a system in a positive energy state there is a probability of induced transitions into negative energy states. Hence if they were excluded this would be a contradiction. The second objection is mathematical. That is, excluding the negative energy states leads to an incomplete set of wave functions. It is not possible to represent an arbitrary function as an expansion in functions of an incomplete set. This situation led Schrödinger into insurmountable difficulties.

Problem: Suppose that for $t < 0$ a particle is in a positive energy state moving in the x direction with spinup in the z direction ($s = +1$). Then at $t = 0$, a constant potential $A = A_z$ ($A_x = A_y = 0$) is turned on and at $t = T$ it is turned off. Find the probability that the particle is in a negative energy state at $t = T$.

Answer:

$$\left.\begin{array}{l}\text{Probability of being in}\\ \text{negative energy state}\\ \text{at } t = T\end{array}\right\} = A^2/(A^2 + m^2) \sin^2[(m^2 + A^2)^{1/2} T]$$

Note that when $E = -m$, $1/\sqrt{F} = \infty$, so the u's apparently blow up. But actually the components of u also vanish when $E = -m$, so that a limiting process is involved. It may be avoided and the correct results obtained simply by omitting $1/\sqrt{F}$ and replacing F by zero and $p_\pm$ by 1 in the components of u.

The positive energy levels form a continuum extending from $E = m$ to $+\infty$, and the negative energies if accepted as such form another continuum from $E = -m$ to $-\infty$. Between $+m$ and $-m$ there are no available energy levels (see Fig. 14-1). Dirac proposed the idea that all the negative energy levels are normally filled. Explanations for the apparent obscurity of such a sea of electrons in negative energy states, if it exists, usually contain a psychological aspect and are not very satisfactory. But, nevertheless, if such a situation is assumed toexist, some of the important consequences are these:

1. Electrons in positive energy states will not normally be observed to make transitions into negative energy states because these states are not available; they are already full.

2. With the sea of electrons in negative energy levels unobservable, a "hole" in it produced by a transition of one of its electrons into a positive energy state should manifest itself. The manifestation of the hole is regarded as a positron and behaves like an electron with a positive charge.

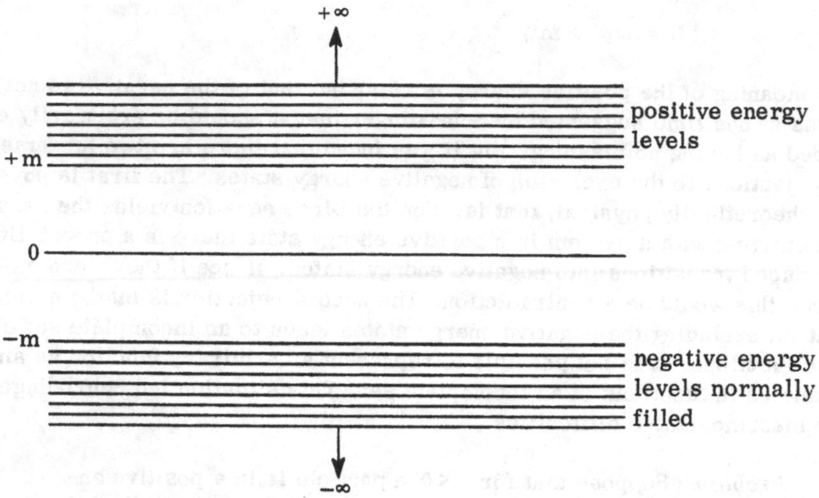

FIG. 14-1

3. The Pauli exclusion principle is implied in order that the negative sea may be full. That is, if any number rather than just one electron could occupy a given state, it would be impossible to fill all the negative energy states. It is in this way that the Dirac theory is sometimes considered as "proof" of the exclusion principle.

Another interpretation of negative energy states has been proposed by the present author. The fundamental idea is that the "negative energy" states represent the states of electrons *moving backward in time*.

In the classical equation of motion

$$m(d^2z_\mu/ds^2) = e(dz_\nu/ds)F_{\mu\nu}$$

reversing the direction of proper time s amounts to the same as reversing the sign of the charge so that the electron moving backward in time would look like a positron moving forward in time.

In elementary quantum mechanics, the total amplitude for an electron to go from x_1,t_1 to x_2,t_2 was computed by summing the amplitudes over all possible trajectories between x_1,t_1 and x_2,t_2, assuming that the trajectories always moved forward in time. These trajectories might appear in one dimension as shown in Fig. 14-2. But with the new point of view, a possible trajectory might be as shown in Fig. 14-3.

Imagining oneself an observer moving along in time in the ordinary way, being conscious only of the present and past, the sequence of events would appear as follows:

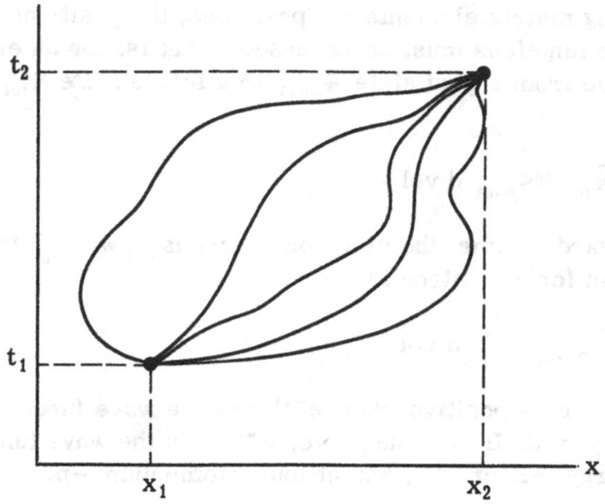

FIG. 14-2

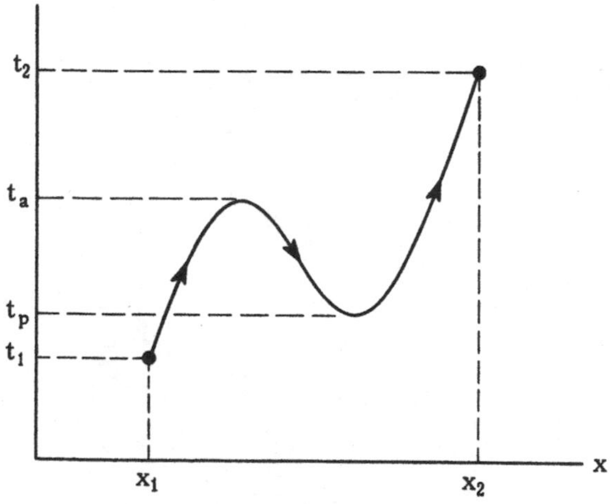

FIG. 14-3

$t_1 \rightarrow t_p$	only the initial electron present
$t_p \rightarrow$	the initial electron still present but somewhere else an electron-positron pair is formed
$t_p \rightarrow t_a$	the initial electron and newly arrived electron and positron are present
$t_a \rightarrow$	the positron meets with the initial electron, both of them annihilating, leaving only the previously created electron
$t_a \rightarrow t_2$	only one electron present

To handle this idea quantum mechanically two rules must be followed:

1. In calculating matrix elements for positrons, the positions of the initial and final wave functions must be reversed. That is, for an electron moving forward in time from a past state Ψ_{past} to a future state Ψ_{fut}, the matrix element is

$$\int \tilde{\Psi}_{fut} M \Psi_{past} \, d\,vol$$

But moving backward in time, the electron proceeds *from* Ψ_{fut} *to* Ψ_{past} so the matrix element for a positron is

$$\int \tilde{\Psi}_{past} M \Psi_{fut} \, d\,vol$$

2. If the energy E is positive, then $e^{-ip\cdot x}$ is the wave function of an electron with energy $p_4 = E$. If E is negative, $e^{-ip\cdot x}$ is the wave function of a positron with energy $-E$ or $|E|$, and of four-momentum $-p$.

Potential Problems in Quantum Electrodynamics

Fifteenth Lecture

PAIR CREATION AND ANNIHILATION

Two possible paths of an electron being scattered between the states Ψ_1 and Ψ_2 were discussed in the last lecture. These are:

Case I. Both Ψ_1, Ψ_2 states of positive energy, interpreted as Ψ_1 electron in "past," Ψ_2 electron in "future." This is electron scattering.

Case II. Both Ψ_1, Ψ_2 states of negative energy interpreted as Ψ_1 positron in "future," Ψ_2 positron in "past." This is positron scattering.

The existence of negative energy states makes two more types of paths possible. These are:

Case III. The Ψ_1 positive energy, Ψ_2 negative energy, interpreted as Ψ_1 in "past," Ψ_2 positron in "past." Both states are in the past, and nothing in the future. This represents pair annihilation.

Case IV. The Ψ_1 negative energy, Ψ_2 positive energy, interpreted as Ψ_1 positron in "future," Ψ_2 electron in "future." This is pair creation.

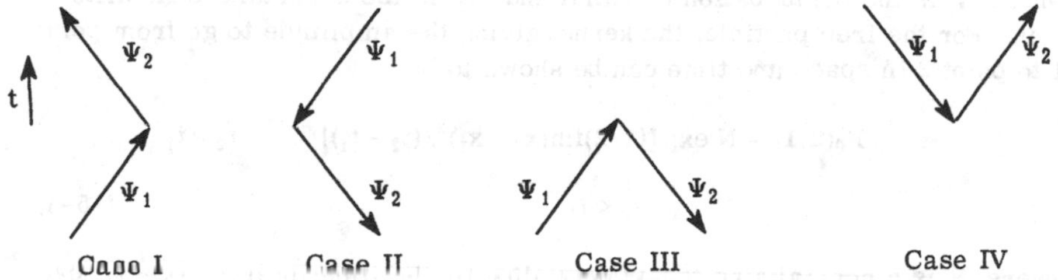

FIG. 15-1

The four cases can be diagrammed as shown in Fig. 15-1. Note that in each diagram the arrows point from Ψ_1 to Ψ_2, although time is increasing upward in all cases. The arrows give the direction of motion of the electron in the present interpretation of negative energy states. In common language, the arrows point toward positive or negative time according to whether $\not{p}$ is positive or negative, that is, whether the state represented is that of an electron or a positron.

CONSERVATION OF ENERGY

Energy relations for the scattering in case I have been established in previous lectures. It can be seen that identical results hold for case II. To show this, recall that in case I, if the electron goes from the energy E_1 to E_2 and if the perturbation potential is taken proportional to $\exp(-i\omega t)$, then this perturbation brings in a positive energy ω. To see this, note that the amplitude for scattering is proportional to

$$\int \exp(-iE_2 t)^* \exp(-i\omega t) \exp(-iE_1 t)\, dt$$

$$= \int \exp[(iE_2 t - i\omega t - iE_1 t)\, dt] \qquad (15\text{-}1)$$

As has been shown, there is a resonance between E_2 and $E_1 + \omega$, so that the only contributing energies are those for which $E_2 \approx E_1 + \omega$. In case II the same integral holds but E_2 and E_1 are negative. A positron goes from an energy (past) of $E_{past} = -E_2$ to an energy (future) of $E_{fut} = -E_1$. With the same perturbation energy, the amplitude is large again only if $E_2 = E_1 + \omega$ or $-E_{past} = -E_{fut} + \omega$, so that $E_{fut} = \omega + E_{past}$; that is, the perturbation carries in a positive energy ω, just as it does for the electron case.

THE PROPAGATION KERNEL

In the nonrelativistic case (Schrödinger equation), the wave equation, including a perturbation potential, is written

$$i\partial\Psi/\partial t = H_0\Psi + V\Psi \qquad (15\text{-}2)$$

where V is the perturbation potential and H_0 is the unperturbed Hamiltonian. For the free particle, the kernel giving the amplitude to go from point 1 to point 2 in space and time can be shown to be

$$K_0(2,1) = N \exp[(1/2)im(x_2 - x_1)^2/(t_2 - t_1)] \qquad t_2 > t_1$$

$$= 0 \qquad t_2 < t_1 \qquad (15\text{-}3)$$

where N is a normalizing factor depending on the time interval $t_2 - t_1$ and the mass of the particle:

$$N = [m/2\pi i(t_2 - t_1)]^{1/2}$$

Note that the kernel is defined to be 0 for $t_2 < t_1$. It can be shown that K_0 satisfies the equation

$$[i\partial/\partial t_2 - H_0(2)]\, K_0(2,1) = i\delta(2,1) \tag{15-4}$$

The propagation kernel $K_V(2,1)$ giving a similar amplitude, but in the presence of the perturbation potential V, must satisfy the equation

$$[i\partial/\partial t_2 - H_0(2) - V(2)]\, K_V(2,1) = i\delta(2,1) \tag{15-5}$$

It can be shown that K_V can be computed from the series

$$K_V(2,1) = K_0(2,1) - i \int K_0(2,3)V(3)K_0(3,1)d^3x_3\, dt_3$$
$$- \int K_0(2,4)V(4)K_0(4,3)V(3)K_0(3,1)d^3x_4\, dt_4\, d^3x_3\, dt_3 + \cdots \tag{15-6}$$

In case the complete Hamiltonian $H = H_0 + V$ is independent of time, and all the stationary states ϕ_n of the system are known, then $K_V(2,1)$ may be obtained from the sum

$$K_V(2,1) = \sum_n \exp[-iE_n(t_2 - t_1)]\, \phi_n(x_2)\phi_n{}^*(x_1) \tag{15-7}$$

The extension of these ideas to the relativistic case (Dirac equation) is straightforward. By choosing a particular form for the Hamiltonian, the Dirac equation can be written

$$i\partial\Psi/\partial t = H\Psi = \boldsymbol{\alpha} \cdot (\mathbf{p} - e\mathbf{A})\Psi + e\phi\,\Psi + m\beta\Psi$$

Defining the propagation kernel as K^A, then the kernel is the solution to the equation

$$[i\partial/\partial t_2 - e\phi_2 - \boldsymbol{\alpha} \cdot (-i\nabla - e\mathbf{A}_2) - m\beta]\, K^A(2,1) = i\beta\delta(2,1) \tag{15-8}$$

The matrix β is inserted in the last term in order that the kernel derived from the Hamiltonian be relativistically invariant. [Note the similarity to the nonrelativistic case, Eq. (15-6).] Multiplying this equation by β, a simpler form results:

$$(i\nabla\!\!\!/_2 - eA\!\!\!/_2 - m)K^A(2,1) = i\delta(2,1) \tag{15-9}$$

The equation for a free particle is obtained simply by letting $A\!\!\!/_2 = 0$, then calling the free-particle kernel K_+,

$$(i\nabla\!\!\!/_2 - m)\, K_+(2,1) = i\delta(2,1) \tag{15-10}$$

The notation K_+ replaces the K_0 of the nonrelativistic case, and Eq. (15-10) replaces Eq. (15-4) as the defining equation.

Just as K_V can be expanded in the series of Eq. (15-6), so K^A can be expanded as

$$K^A(2,1) = K_+(2,1) - i \int K_+(2,3) e A\!\!\!/(3) K_+(3,1)\, d\tau_3$$
$$- \int K_+(2,3) e A\!\!\!/(3) K_+(3,4) e A\!\!\!/(4) K_+(4,1) d\tau_3\, d\tau_4 + \cdots \tag{15-11}$$

Note that the kernel is now a four-by-four matrix, so that all components of Ψ can be determined. Since this is true, the order of the terms in Eq. (15-11) is important. The element of integration is actually an element of volume in four-space,

$$d\tau = dx_1\, dx_2\, dx_3\, dx_4$$

The potential, $-ieA\!\!\!/(1)$ can be interpreted as the amplitude per cubic centimeter per second for the particle to be scattered once at the point (1). Thus the interpretation of Eq. (15-11) is completely analogous to that of Eq. (15-6).

Problem: Show that K^A as defined by Eq. (15-11) is consistent with Eqs. (15-8) and (15-9).

On the nonrelativistic case, the paths along which the particle reversed its motion in time are excluded. In the present case this is no longer true. The existence and interpretation of the negative energy eigenvalues of the Dirac equation allows the interpretation and inclusion of such paths.

Taking $t_4 > t_3$ implies the existence of virtual pairs. The section from t_4 to t_3 represents the motion of a positron (see Fig. 15-2).

In a time-stationary field, if the wave functions ϕ_n are known for all the states of the system, then $K_+{}^A$ may be defined by

$$K_+{}^A(2,1) = \sum_{\text{pos. energies}} \exp[-iE_n(t_2 - t_1)]\, \phi_n(x_2)\tilde{\phi}_n(x_1) \qquad t_2 > t_1$$

$$= - \sum_{\text{neg. energies}} \exp[-iE_n(t_2 - t_1)]\, \phi_n(x_2)\tilde{\phi}_n(x_1) \qquad t_2 < t_1 \tag{15-12}$$

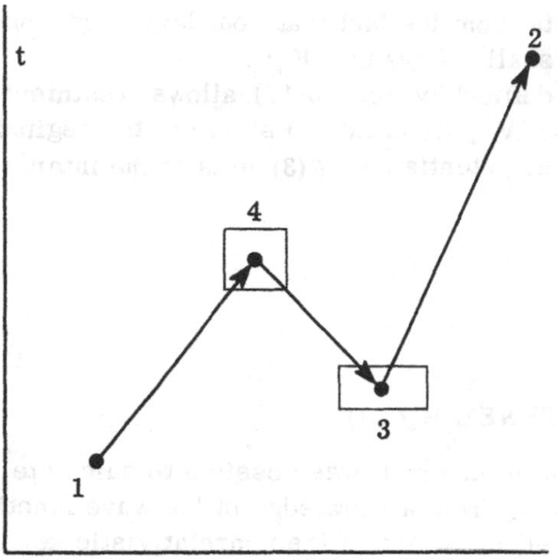

FIG. 15-2

Another solution of Eq. (15-9) is

$$K_0^A(2,1) = \sum_{\text{pos. energies}} \exp[-iE_n(t_2-t_1)]\phi_n(x_2)\tilde{\phi}_n(x_1)$$

$$+ \sum_{\text{neg. energies}} \exp[-iE_n(t_2-t_1)]\phi_n(x_2)\tilde{\phi}_n(x_1) \qquad t_2 > t_1$$

$$= 0 \qquad t_2 < t_1 \qquad (15\text{-}13)$$

Equation (15-13) has an interpretation consistent with the positron interpretation of negative energy states. Thus when the timing is "ordinary" ($t_2 > t_1$), an electron is present, and only positive energy states contribute. When the timing is "reversed" ($t_2 < t_1$), a positron is present, and only negative energy states contribute. On the other hand, Eq. (15-13) does not have so satisfactory an interpretation. Although the kernel K_0^A defined by Eq. (15-13) is also a satisfactory mathematical solution of Eq. (15-9) (as shown below), the interpretation of Eq. (15-13) requires the idea of an electron in a negative energy state.

To show that both kernels are solutions of the same inhomogeneous equation, note that their difference is

$$\sum_{\text{neg. energies}} \exp(iE_n t_1)\exp(-iE_n t_2)\phi_n(x_2)\tilde{\phi}_n(x_1)$$

for *all* t_2. This is, term by term, a solution of the homogeneous equation [i.e., Eq. (15-9) with zero right-hand side]. The possibility that two such solutions exist results from the fact that boundary conditions have not been definitely fixed. We shall always use K_+^A.

The kernel K_+^A, defined by Eq. (15-12), allows treatment of case III (pair annihilation) and case IV (pair creation) shown at the beginning of this lecture. In each case, the potential, $-ie\not{A}(3)$, acts at the intersection of positron and electron paths.

Sixteenth Lecture

USE OF THE KERNEL $K_+(2,1)$

In the nonrelativistic theory it was possible to calculate the wave function at a point x_2 at time t_2 from a knowledge of the wave function at an earlier time t_1 (see Fig. 16-1) by means of the nonrelativistic kernel $K_0(\mathbf{x}_2,t_2;\ \mathbf{x}_1,t_1)$,

$$\Psi(\mathbf{x}_2,t_2) = \int K_0(\mathbf{x}_2,t_2;\ \mathbf{x}_1,t_1)\Psi(\mathbf{x}_1,t_1)\ d^3x_1$$

It might be expected that a relativistic generalization of this would be

$$\Psi(\mathbf{x}_2,t_2) = \int K_+(\mathbf{x}_2,t_2;\ \mathbf{x}_1,t_1)\gamma_t\Psi(\mathbf{x}_1,t_1)\ d^3x_1$$

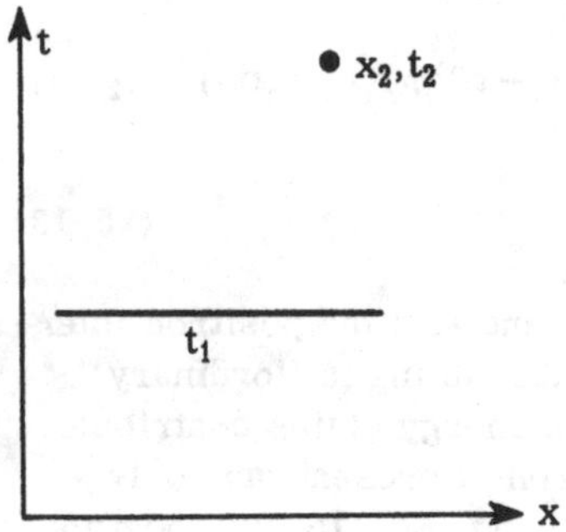

FIG. 16-1

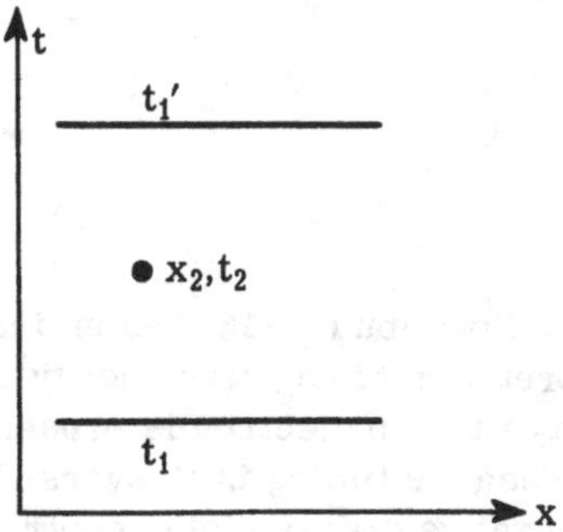

FIG. 16-2

This turns out to be incorrect, however. It is not sufficient, in the relativistic case, to know just the wave function at an earlier time only because $K_+(2,1)$ is not zero for $t_2 < t_1$. When the kernel is defined in this manner (Lecture 15), the wave function at $\mathbf{x}_2,t_2$ (see Fig. 16-2) is given by

$$\Psi(\mathbf{x}_2,t_2) = \int K_+(\mathbf{x}_2,t_2;\ \mathbf{x}_1,t_1)\gamma_t\Psi(\mathbf{x}_1,t_1)\,d^3x_1 - \int K_+(\mathbf{x}_2,t_2;\ \mathbf{x}_1,t_1')\gamma_t\Psi(\mathbf{x}_1,t_1')\,d^3x_1 \qquad t_1 < t_2 < t_1' \tag{16-1}$$

The first term is the contribution from positive energy states at earlier times and the second term is the contribution from negative energy states at later times. This expression can be generalized to state that it is necessary to know $\Psi(x_1 t_1)$ on a four-dimensional surface, surrounding the point x_2, t_2 (see Fig. 16-3):

$$\Psi(x_2,t_2) = \int K_+(2,1)\not{N}(1)\Psi(1)\, d^4x_1 \tag{16-2}$$

where $\not{N}_\mu$ is the four-vector normal to the surface that encloses x_2, t_2.

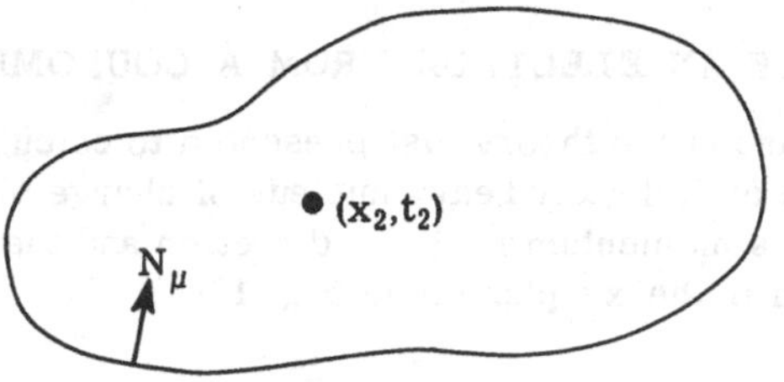

FIG. 16-3

TRANSITION PROBABILITY

The amplitude to go from a state f to a state g under the action of a potential $\not{A}$ is given by an expression similar to that in nonrelativistic theory,

$$a_{21} = \int\int \tilde{g}(2)\beta K_+^{A}(2,1)\beta f(1) d^3x_1\, d^3x_2 \tag{16-3}$$

Using the expansion of $K_+^{A}(2,1)$ in terms of $K_+(2,1)$, Eq. (15-12), and assuming that the amplitude for transition from state f to state g as a free particle is zero (f and g orthogonal states), the first-order amplitude for transition (Born approximation) is

$$a_{21} = -i \int \tilde{g}(2)\, \beta \int K_+(2,3) e\not{A}(3) K_+(3,1)\beta f(1)\, d\tau_3\, d^3x_1\, d^3x_2$$

It is convenient to let

$$f(3) = \int K_+(3,1)\beta f(1)\, d^3x_1$$

$$\tilde{g}(3) = \int \tilde{g}(2)\beta K_+(2,3)\, d^3x_2$$

These state that the particle has the free-particle wave function f just prior to scattering and the free-particle wave function g just after scattering, and that it eliminates any computation of the motion as a free particle. The amplitude for transition, to first order, may be written

$$-i\int \tilde{g}(3) e \not{A}(3) f(3)\, d\tau_3 \tag{16-4}$$

($d\tau_3$ signifies integration over time as well as space). The second-order term would be written

$$-(1/2)\int\int g(4) e \not{A}(4) K_+(4,3) e \not{A}(3) f(3)\, d\tau_3\, d\tau_4$$

If f(3) is a negative energy state, then it represents a positron of the future instead of an electron of the past and the process described by this amplitude is that of pair production.

SCATTERING OF AN ELECTRON FROM A COULOMB POTENTIAL

We shall make use of the theory just presented to calculate the scattering of an electron from an infinitely heavy nucleus of charge Ze. Suppose the incident electron has momentum in the x direction and the scattered electron has momentum in the xy plane (see Fig. 16-4):

$$\not{p}_1 = \gamma_t E_1 - \gamma_x p_{1x}$$

$$\not{p}_2 = \gamma_t E_z - \gamma_x p_{2x} - \gamma_y p_{2y}$$

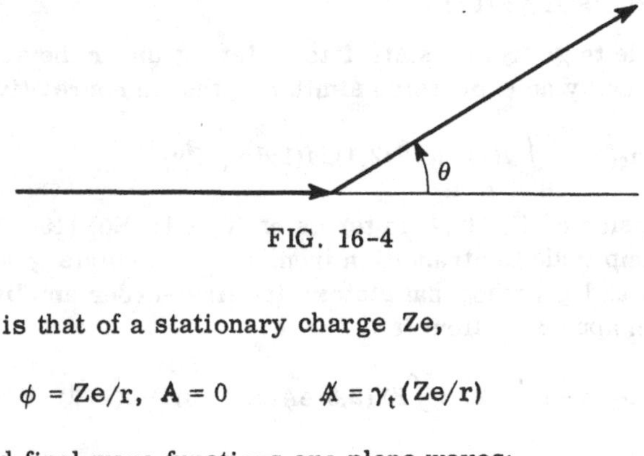

FIG. 16-4

The potential is that of a stationary charge Ze,

$$\phi = Ze/r, \quad \mathbf{A} = 0 \qquad \not{A} = \gamma_t (Ze/r)$$

The initial and final wave functions are plane waves:

$$f(1) = u_1 e^{-ip_1\cdot x} \qquad g(2) = u_2 e^{-ip_2\cdot x}$$ (four-component wave function)

Thus, by Eq. (16-4), the first-order amplitude for transition from state f to state g (momentum p_1 to momentum p_2) is

$$M = -i\int \tilde{u}_2 e^{ip_2\cdot x} (Ze^2/r)\gamma_t u_1 e^{-ip_1\cdot x}\, d^3x\, dt$$

Separating space and time dependence in the wave functions, this becomes

$$M = -i(\tilde{u}_2\gamma_t u_1)\left[\int e^{-ip_2\cdot x}(Ze^2/r)e^{ip_1\cdot x}\,d^3x\right]\left[\int_0^T e^{iE_2t}\,e^{-iE_1t}\,dt\right]$$

The first integral is just V(**Q**), a three-dimensional Fourier transform of the potential, which was evaluated in nonrelativistic scattering theory:

$$M = -i(\tilde{u}_2\gamma_t u_1)[V(\mathbf{Q})]\left\{\frac{\exp[i(E_2-E_1)t]-1}{i(E_2-E_1)}\right\} \qquad (16\text{-}5)$$

$$V(\mathbf{Q}) = 4\pi Ze^2/Q^2 \qquad \mathbf{Q} = \mathbf{p}_1 - \mathbf{p}_2$$

The probability of transition per second is given by

$$\text{Trans. prob./sec} = 2\pi(\Pi N)^{-1}|M|^2 \times (\text{density of final states}) \qquad (16\text{-}6)$$

This is a result from time-dependent perturbation theory, the only new factor is a normalizing factor $(\Pi N)^{-1}$ which takes account for the fact that the wave functions are not normalized to unity per unit volume. The ΠN is a product of factors N one for each wave function, or particle in the initial state, and one for each final wave function,

$$N = (\tilde{u}\gamma_t u) \qquad (16\text{-}7)$$

for each particle in question. In our normalization, then $N = 2E$.

The reason for this factor is that wave functions are normalized to

$$(\tilde{u}u) = 2m \qquad \text{or} \qquad (\tilde{u}\gamma_t u) = 2E$$

where, as in the computation of transition probability, they should be normalized in the conventional nonrelativistic manner $\Psi^*\Psi = 1$ or $(\tilde{u}\gamma_t u) = 1$ (so $N = 1$ for that case).

The matrix element M, as calculated in this manner, is relativistically invariant and in the future the chief interest will be in M. The transition probability, knowing M, can be computed from Eq. (16-6).

Density of States, Cross Section. For the electron scattering problem under consideration,

$$M = -i(\tilde{u}_2\gamma_t u_1)(4\pi Ze^2/Q^2)$$

so the transition probability is

$$\begin{matrix}\text{Trans.}\\ \text{prob./sec}\end{matrix} = \frac{2\pi}{(2E_1)(2E_2)}|(\tilde{u}_2\gamma_t u_1)|^2\left|\frac{4\pi Ze^2}{Q^2}\right|^2\frac{E_2p_2\,d\Omega}{(2\pi)^3} \qquad (16\text{-}8)$$

where the density of final states has been obtained in the following manner:

$$\text{Density of states} = \frac{d^3 p_2}{(2\pi)^3\, dE_2} = \frac{p_2^2\, dp_2\, d\Omega}{(2\pi)^3\, dE_2} \qquad \hbar = 1$$

but $E_2^2 = p_2^2 + m^2$, so $dp_2/dE_2 = E_2/p_2$ and

$$\text{Density of states} = E_2 p_2 d\Omega/(2\pi)^3$$

When the incoming plane wave is normalized to one particle per cubic centimeter, the cross section is given in terms of the transition probability per second† as

$$\text{Trans. prob./sec} = \sigma v_1 = \sigma (p_1/E_1)$$

or

$$\sigma = (E_1/p_1) \times (\text{trans. prob./sec})$$

The essential difference between the relativistic treatment of scattering and the nonrelativistic treatment is contained in the matrix element $(\tilde{u}_2\gamma_t u_1)$. From Table 13-1, for a particle moving in the xy plane and $s_1 = +1$, $s_2 = +1$,

$$|(\tilde{u}_2\gamma_t u_1)|^2 = 1/F_1F_2|F_2F_1 + p_{1+}p_{2-}|^2$$

where

$$F_1 = F_2 = E + m$$

$[E_1 = E_2$, conservation of energy, follows from the nature of the time integral in Eq. (16-5)], and

$$p_{1+} = p$$

$$p_{2-} = pe^{-i\theta}$$

(magnitude of final momentum equal to magnitude of initial momentum follows from $E_1 = E_2$).

Thus

$$\dagger\, p_1 = \frac{mv_1}{(1-v_1^2)^{1/2}} \rightarrow p_1^2 = \frac{m^2v_1^2}{1-v_1^2} \rightarrow p_1^2 = (m^2+p_1^2)v_1^2 = E_1^2v_1^2$$

Therefore, $v_1 = p_1/E_1$.

$$|(\tilde{u}_2\gamma_t u_1)|^2 = (E+m)^{-2}\,|(E+m)^2 + p^2e^{-i\theta}|^2$$
$$= (E+m)^{-2}\,\{4E^2(E+m)^2[1-(p^2/E^2)\sin^2(\theta/2)]\}$$
$$= (2E)^2\,[1 - v^2\sin^2(\theta/2)]$$

When $s_1 = +1$, $s_2 = -1$ or $s_1 = -1$, $s_2 = +1$, the matrix element of γ_t is zero. When $s_1 = -1$, $s_2 = -1$, the absolute value of the matrix element is the same as for $s_1 = +1$, $s_2 = +1$. Thus spin does not change in scattering (in Born approximation) and the cross section is independent of spin,

$$\sigma = (4Z^2e^4E^2/Q^4)d\Omega\,[1 - v^2\sin^2(\theta/2)] \qquad Q = 2p\sin(\theta/2)$$

The criterion for validity of the Born approximation, used in obtaining this result, is $Ze^2/\hbar v \ll 1$. In the extreme relativistic limit $v \approx c$. This becomes $Z \ll 137$. Just as for the nonrelativistic case, the scattering can actually be calculated exactly (correct to all orders in the potential) for the Coulomb potential. This exact solution of the Dirac equation involves hypergeometric functions. It was first worked out by Mott and is called Mott scattering. For moderate energies (200 kev) there is some probability for change in spin. Polarized electrons could be produced in this manner.

Problems: (1) Calculate the Rutherford scattering law for the Klein-Gordon equation (particle with no spin). Result: same formula as just given with $1 - v^2\sin^2(\theta/2)$ replaced by 1.

(2) Show that this scattering formula is also correct for positrons (use positron states in calculating matrix element).

Seventeenth Lecture

CALCULATION OF THE PROPAGATION KERNEL FOR A FREE PARTICLE

As shown in a previous lecture, the propagation kernel, when there is no perturbing potential and the Hamiltonian of the system is constant in time, is

$$K_+(2,1) = \sum_{+n}\phi_n(\mathbf{x}_2)\,\tilde{\phi}_n(\mathbf{x}_1)\exp[-iE_n(t_2-t_1)] \qquad t_2 > t_1$$
$$= -\sum_{-n}\phi_n(\mathbf{x}_2)\,\tilde{\phi}_n(\mathbf{x}_1)\exp[-iE_n(t_2-t_1)] \qquad t_2 < t_1$$

For a free particle, the eigenfunctions ϕ_n are

$$u_p\exp(i\mathbf{p}\cdot\mathbf{x})$$

and the sum over n becomes an integral over p. The u_p is the spinor corresponding to momentum p, positive or negative energy and spin up or down, as appropriate. Then the propagation kernel for a free particle is, for $t_2 > t_1$,

$$K_+(2,1) = \sum_{\text{spins}} \int \frac{d^3p}{(2\pi)^3} \frac{1}{2E_p} u_p \tilde{u}_p \exp[i\mathbf{p}\cdot(\mathbf{x}_2 - \mathbf{x}_1)]$$

$$\times \exp[-iE_p(t_2 - t_1)]$$

for $E_p = +(p^2 + m^2)^{1/2}$. The factor $1/(2\pi)^3$ is the density of states per unit volume of momentum space per cubic centimeter. The factor $1/2E_p$ arises from the normalization $u\tilde{u} = 2m$ or $u\gamma_t\tilde{u} = 2E_p$ used here. The u_p are those for positive energy. For negative energy $E_p = -(p^2 + m^2)^{1/2}$, the u_p are changed accordingly and $K_+(2,1)$ becomes, for $t_2 < t_1$,

$$K_+(2,1) = -\sum_{\text{spins}} \int \frac{d^3p}{(2\pi)^3} \frac{1}{2E_p} u_p \tilde{u}_p \exp[i\mathbf{p}\cdot(\mathbf{x}_2 - \mathbf{x}_1)]$$

$$\times \exp[iE_p(t_2 - t_1)]$$

The calculation will be made first for the case of $t_2 > t_1$. We first calculate $u_p\tilde{u}_p$ for positive energy, and p in the xy plane and spin up. Under these conditions

$$u_p = \begin{pmatrix} E + m \\ 0 \\ 0 \\ p_x + ip_y \end{pmatrix} \frac{1}{(E+m)^{1/2}}$$

$$\tilde{u}_p = \begin{pmatrix} E+m & 0 & 0 & -p_x + ip_y \end{pmatrix} \frac{1}{(E+m)^{1/2}}$$

Note that $u_p\tilde{u}_p$ is the opposite order to that usually encountered so that the product is a matrix, not a scalar. That is,

$$u_p\tilde{u}_p = \begin{pmatrix} (E+m)^2 & 0 & 0 & (E+m)(-p_x+ip_y) \\ 0 & 0 & 0 & 0 \\ 0 & 0 & 0 & 0 \\ (E+m)(p_x+ip_y) & 0 & 0 & (p_x+ip_y)(-p_x+ip_y) \end{pmatrix}$$

$$\times\ 1/(E+m)$$

by the usual rules for matrix multiplication. But

$$(p_x + ip_y)(-p_x + ip_y) = -p^2 = -E^2 + m^2$$

and the matrix becomes

$$u_p\tilde{u}_p = \begin{pmatrix} E+m & 0 & 0 & -p_x+ip_y \\ 0 & 0 & 0 & 0 \\ 0 & 0 & 0 & 0 \\ p_x+ip_y & 0 & 0 & -E+m \end{pmatrix} \quad \text{(spin up)}$$

By the same process, the result in the spin down case is

$$u_p = \begin{pmatrix} 0 \\ E+m \\ p_x - ip_y \\ 0 \end{pmatrix} \frac{1}{(E+m)^{1/2}}$$

$$\tilde{u}_p = \left(0 \quad E-m \quad -p_x - ip_y \quad 0 \right) \frac{1}{(E+m)^{1/2}}$$

$$u_p\tilde{u}_p = \begin{pmatrix} 0 & 0 & 0 & 0 \\ 0 & E+m & -p_x-ip_y & 0 \\ 0 & p_x-ip_y & -E+m & 0 \\ 0 & 0 & 0 & 0 \end{pmatrix} \quad \text{(spin down)}$$

It may be verified easily that the sum of these matrices for spin up and spin down is represented by

$$E\gamma_t - p_x\gamma_x - p_y\gamma_y + m$$

In the general case when **p** is in any direction, it is clear that the only change is an additional term $-p_z\gamma_z$. So, in general,

$$(u_p\tilde{u})_{\text{spin up}} + (u_p\tilde{u}_p)_{\text{spin down}} = E\gamma_t - \mathbf{p}\cdot\boldsymbol{\gamma} + m = \not{p} + m$$

The sign of the energy was not used in obtaining this result so it is the same for either sign.

Now put $t_2 - t_1 = t$ and $\mathbf{x}_2 - \mathbf{x}_1 = \mathbf{x}$. For $t > 0$, the propagation kernel becomes

$$K_+(2,1) = \int (E_p\gamma_t - \mathbf{p}\cdot\boldsymbol{\gamma} + m)[d^3\mathbf{p}/(2\pi)^3](1/2E_p)$$
$$\times \exp[-i(E_p t - \mathbf{p}\cdot\mathbf{x})]$$

The appearance of p in the form $E_p = (p^2 + m^2)^{1/2}$ in the time part of the exponential makes this a difficult integral. Note that it may also be written in the form

$$K_+(2,1) = \left(i\gamma_t \frac{\partial}{\partial t} + i\gamma_x \frac{\partial}{\partial x} + i\gamma_y \frac{\partial}{\partial y} + i\gamma_z \frac{\partial}{\partial z} + m\right)$$
$$\times \int \frac{d^3\mathbf{p}}{(2\pi)^3 2E_p} \exp[-i(E_p t - \mathbf{p}\cdot\mathbf{x})]$$
$$= i(i\not{\nabla} + m)\, I_+(t,\mathbf{x})$$

where

$$I_+(t,\mathbf{x}) = -i \int \frac{d^3\mathbf{p}}{(2\pi)^3 2E_p} \exp[-i(E_p t - \mathbf{p}\cdot\mathbf{x})]$$

In this form only one integral instead of four need be done. It may be verified as an exercise that for $t < 0$ the result is the same except that the sign of t is changed, so that putting $|t|$ in place of t in the formula for $I_+(t,\mathbf{x})$ makes it good for all t.

This integral has been carried out with the following result:

$$I_+(t,\mathbf{x}) = -(4\pi)^{-1}\delta(s^2) + (m/8\pi s)H_1^{(2)}(ms)$$

where $s = +(t^2 - x^2)^{1/2}$ for $t > x$, and $-i(x^2 - t^2)^{1/2}$ for $t < x$. $\delta(s^2)$ is a delta function and $H_1^{(2)}(ms)$ is a Hankel function.† Another expression for the foregoing is

$$I_+(t,\mathbf{x}) = -(1/8\pi^2)\int_0^\infty d\alpha \exp\{-(i/2)[(m^2/\alpha) + \alpha(t^2 - x^2)]\}$$

Both of these forms are too complicated to be of much practical use. It will be shown shortly that a tremendous simplification results from transformation to momentum representation.

Note that $I_+(t,\mathbf{x})$ actually depends only on $|\mathbf{x}|$, not on its direction. In the time-space diagram (Fig. 17-1) the space axis represents $|\mathbf{x}|$ and the diagonal lines represent the surface of a light cone including the t axis, that is, the accessible region of $t - |\mathbf{x}|$ space in the ordinary sense. It can be shown that the asymptotic form of $I_+(t,\mathbf{x})$ for large s is proportional to e^{-ims}. When one's region of accessibility is limited to the inside of the light cone, large s implies $t^2 \gg |\mathbf{x}|^2$, so that the region of the asymptotic approximation lies roughly within the dotted cone around the t axis and is

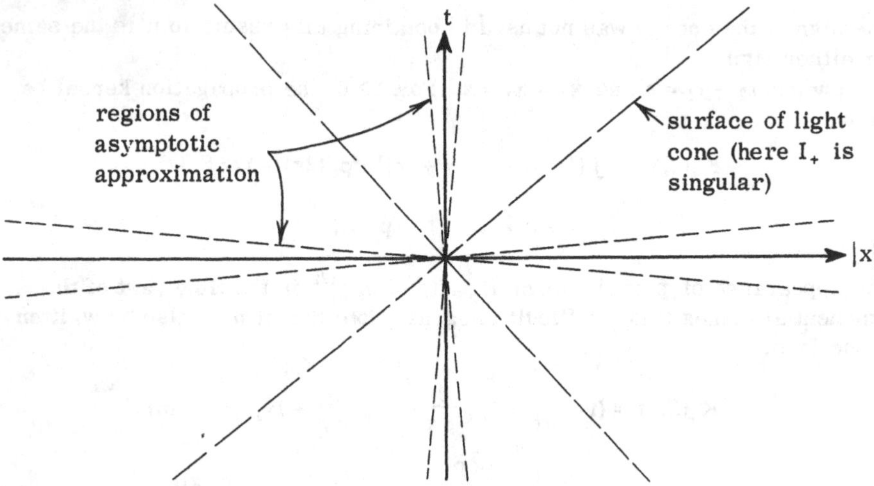

FIG. 17-1

† See Phys. Rev., 76, 749 (1949); included in this volume.

$$I_+(t,\mathbf{x}) \to e^{-ims} \approx \exp\{-im[t-(x^2/2t)]\} \approx e^{-imt}$$

The first form is seen to be essentially the same as the propagation kernel for a free particle used in nonrelativistic theory. If, as in the new theory, possible "trajectories" are not limited to regions within the light cone, another region included in this asymptotic approximation is that within the dotted cone along the $|\mathbf{x}|$ axis where large s implies $|x|^2 \gg t^2$. Hence

$$I_+(t,\mathbf{x}) \to e^{-ims} = \exp[-im(x^2-t^2)^{1/2}] \approx e^{-m|\mathbf{x}|}$$

It is seen that the distance along $|\mathbf{x}|$ in which this becomes small is roughly the Compton wavelength (recall that $m \to mc/\hbar$ when it represents a length^{-1} as here), so that in reality not much of the $t - |\mathbf{x}|$ space outside the light cone is accessible.

The transformation to momentum representation will now be made. This is facilitated by use of the integral formula

$$\lim_{\epsilon \to 0} \int_{-\infty}^{\infty} dp_4 \frac{\exp(-ip_4 t)}{p_4^2 - E_p^2 + i\epsilon} = -\frac{\pi i}{E_p} \exp(-iE_p|t|)$$

The $i\epsilon$ term in the denominator is introduced solely to ensure passage around the proper side of the singularities at $p_4^2 = E_p^2$ along the path of integration. Passage on the wrong side will reverse the sign in the exponential on the right.

Problem: Work out the integral above by contour integration or otherwise.

Using the integral relation above, $I_+(t,\mathbf{x})$ becomes

$$I_+(t,\mathbf{x}) = \int \frac{d^3p}{(2\pi)^4}\, dp_4\, \frac{\exp(-ip_4 t)\exp(+i\mathbf{p}\cdot\mathbf{x})}{p_4^2 - E_p^2 + i\epsilon}$$

But $E_p^2 = \mathbf{p}^2 + m^2$ so this is

$$I_+(t,\mathbf{x}) = \int \frac{d^4p}{(2\pi)^4} \frac{\exp[-i(p\cdot x)]}{p^2 - m^2 + i\epsilon}$$

where p is now a four-vector so that $d^4p = dp_4\, dp_1\, dp_2\, dp_3$, and $p^2 = p_\mu p_\mu$. Hereafter the $i\epsilon$ term will be omitted. Its effect can be included simply by imagining that m has an infinitesimal negative imaginary part. In this form the transformation to momentum representation is easily accomplished as follows (we actually take Fourier transform of both space and time, so this is really a momentum-energy representation):

$$i_+(p) = \int I_+(t,x) \exp[+i(p \cdot x)]\, d^4x$$

$$= \int \frac{d^4\xi\, d^4x}{(2\pi)^4} \frac{\exp[-i(\xi - p) \cdot x]}{\xi^2 - m^2}$$

where the dummy variable ξ has been substituted for p in the p integral. But

$$\int_{-\infty}^{\infty} \exp[-i(\xi - p) \cdot x]\, d^4x = (2\pi)^4 \delta(\xi - p)$$

Hence the ξ integration gives the result

$$i_+(p) = 1/(p^2 - m^2)$$

Finally, applying the operator $i(i\nabla\!\!\!/ + m)$ to $I_+(t,x)$ gives the propagation kernel (here $x = x_2 - x_1$)

$$K_+(2,1) = i(i\nabla\!\!\!/ + m) I_+(t,x) = i \int \frac{d^4p}{(2\pi)^4} (i\nabla\!\!\!/ + m) \frac{\exp[-i(p \cdot x)]}{p^2 - m^2}$$

$$= i \int \frac{d^4p}{(2\pi)^4} \frac{p\!\!\!/ + m}{p^2 - m^2} \exp[-i(p \cdot x)]$$

recalling that $i\nabla\!\!\!/$ operating on $\exp[-i(p \cdot x)]$ is the same as multiplying by $p\!\!\!/$. From the identity

$$\frac{1}{p\!\!\!/ - m} = \frac{1}{p\!\!\!/ - m} \frac{p\!\!\!/ + m}{p\!\!\!/ + m} = \frac{p\!\!\!/ + m}{p^2 - m^2}$$

the kernel can also be written

$$K_+(2,1) = i \int \frac{d^4p}{(2\pi)^4} \frac{\exp[-i(p \cdot x)]}{p\!\!\!/ - m}$$

By the same process used for $I_+(t,x)$, the transform of $K_+(2,1)$ in momentum representation is seen to be

$$k(p) = \int K_+(2,1) \exp[+i(p \cdot x)]\, d^4x = i[1/(p\!\!\!/ - m)]$$

This is the result which was sought.

Actually this transformation could have been obtained in an elegant manner. For K(2,1) is the Green's function of $(i\nabla\!\!\!/ - m)$, that is,

$$(i\nabla\!\!\!/ - m) K(2,1) = i\delta(2,1) \qquad (17\text{-}1)$$

and it is known that $i\nabla\!\!\!/$ is $p\!\!\!/$ in momentum representation and $\delta(2,1)$ is unity.

Therefore the transform of this equation can be written down immediately:

$$(\not{p} - m)k(p) = i$$

or

$$k(p) = i/(\not{p} - m) \tag{17-2}$$

as before.

The fact that Eq. (17-1) for K(2,1) has more than one solution is reflected in Eq. (17-2) in the fact that $(\not{p} - m)^{-1}$ is singular if $p^2 = m^2$. We shall have to say just how we are to handle poles arising from this source in integrals. The rule that selects the particular form we want is that m be considered as having an infinitesimal negative imaginary part.

Eighteenth Lecture

MOMENTUM REPRESENTATION

Since the propagation kernel for a free particle is so simply expressed in momentum representation,

$$k(p) = i/(\not{p} - m)$$

it will be convenient to convert all our equations to this representation. It is especially useful for problems involving free, fast, moving particles. This requires four-dimensional Fourier transforms. To convert the potential, define

$$\not{a}(q) = \int \not{A}(x)\exp(iq \cdot x)\, d^4x \tag{18-1}$$

Then the inverse transform is

$$\not{A}(x) = (1/2\pi)^4 \int \not{a}(q) \exp(-iq \cdot x)\, d^4q \tag{18-2}$$

The function a(q) is interpreted as the amplitude that the potential contains the momentum (q). As an example, consider the Coulomb potential, given by $\mathbf{A} = 0$, $\varphi = Ze/r$.

Substituting into Eq. (18-1) gives

$$\not{a}(q) = 4\pi Ze/(\mathbf{Q} \cdot \mathbf{Q})\delta(q_4)\gamma_t$$

Here the vector $\mathbf{Q}$ is the space part of the momentum. The delta function $\delta(q_4)$ arises from the time dependence of $\not{A}(x)$.

Matrix Elements. An advantage of momentum representation is the simplicity of computing matrix elements. Recall that in space representation the first-order perturbation matrix element is given by the integral

$$M = -i \int \tilde{g}(2) e \not{A}(2) f(1)\, d\tau_2$$

For the free particle, this becomes

$$M = -i \int \tilde{u}_2 \exp(ip_2 \cdot x_2) e \not{A}(2) u_1 \exp(-ip_1 \cdot x_1)\, d\tau_2 \qquad (18\text{-}3)$$

In momentum representation, this is simply

$$M = i(\tilde{u}_2 e \not{a}(q) u_1) \qquad (18\text{-}3')$$

where $\not{q}$ is defined analogously to the three-vector q,

$$\not{q} = \not{p}_2 - \not{p}_1$$

The second-order matrix element in space representation is given by

$$-\int\int \tilde{g}(2) e \not{A}(2) \overline{K}_+(2,1) e \not{A}(1) f(1)\, d\tau_1\, d\tau_2$$

Substituting for a free particle and also expressing the potential functions as their Fourier transforms by means of Eq. (18-2), this becomes

$$-\int\int\int\int \tilde{u}_2 \exp(ip_2 \cdot x_2) e \not{a}(q_2) \exp(-iq_2 \cdot x_2) K_+(2,1) e \not{a}(q_1)$$
$$\times \exp(-iq_1 \cdot x_1)\, u_1 \exp(-ip_1 \cdot x_1) d\tau_1\, d\tau_2 \cdot d^4q_1/(2\pi)^4$$
$$\times\, d^4q_2/(2\pi)^4 \qquad (18\text{-}4)$$

If Eq. (18-2) is used for $K_+(2,1)$, this kernel can be written

$$K_+(2,1) = \int i/(\not{p} - m) \exp[-ip \cdot (x_2 - x_1)]\, d^4p/(2\pi)^4$$

Writing the factors that depend on τ_1, this part of the integral is

$$\int \exp(ip \cdot x_1) \exp(-iq_1 \cdot x_1) \exp(-ip_1 \cdot x_1)\, d\tau_1$$
$$= (2\pi)^4 \delta^4(p - q_1 - p_1) \qquad (18\text{-}5)$$

where the function $\delta^4(x)$ is to be interpreted as $\delta(t_1)\delta(x_2)\delta(y_3)\delta(z_4)$. Then the integral over τ_1 is zero for all $\not{p}$ except $\not{p} = \not{p}_1 + \not{q}_1$. So the integral over p reduces Eq. (18-4) to

$$-\iiiint \tilde{u}_2 \exp(ip_2 \cdot x_2) e\not{a}(q_2) \exp(-ip_2 \cdot x_2) \exp[-i(p_1 + q_1) \cdot x_2]$$
$$\times\, i(\not{p}_1 + \not{q}_1 - m)^{-1} e\not{a}(q_1) u_1 \; d\tau_2 \; d^4q_1/(2\pi)^4 \; d^4q_2/(2\pi)^4$$

Integrating over τ_2 results in another δ function [similar to Eq. (18-5)], which differs from zero only when

$$\not{p}_2 - \not{q}_2 = \not{p}_1 + \not{q}_1$$

Then integrating over d^4q_2 gives finally

$$(-i^2)i \int \tilde{u}_2 e\not{a}(q_2)(\not{p}_1 + \not{q}_1 - m)^{-1} e\not{a}(q_1) u_1 \; d^4q_1/(2\pi)^4 \qquad (18\text{-}6)$$

These results can be written down immediately by inspection of a diagram of the interaction (see Fig. 18-1). The electron enters the region at 1 with

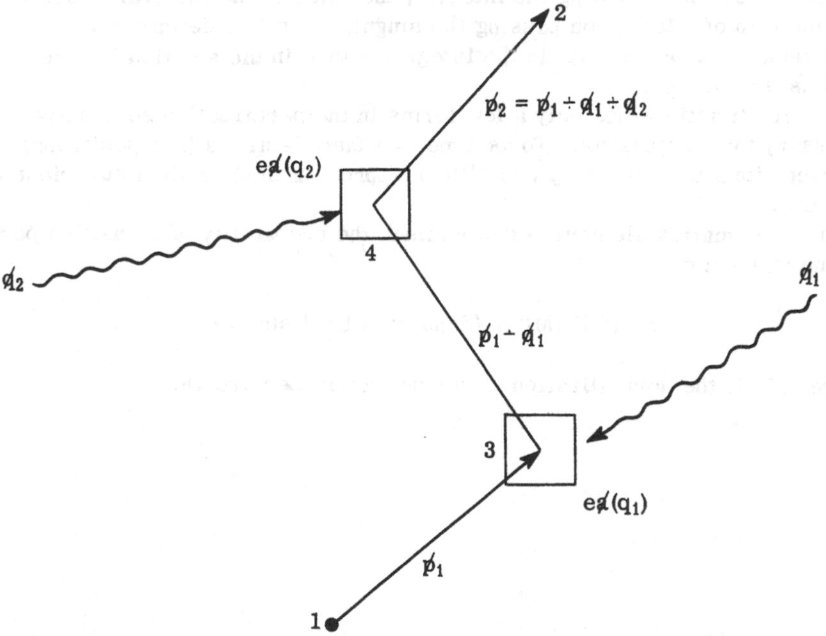

FIG. 18-1

wave function u_1 and moves from 1 to 3 as a free particle of momentum $\not{p}_1$. At point 3, it is scattered by a photon of momentum $\not{q}_1$ [under the action of the potential $-ie\not{a}(q_1)$]. Having absorbed the momentum of the photon it then moves from 3 to 4 as a free particle of momentum $\not{p}_1 + \not{q}_1$ by conservation of momentum. At point 4, it is scattered by a second photon of momentum $\not{q}_2$ [under the action of the potential $-ie\not{a}(q_2)$ absorbing the additional momentum $\not{q}_2$)]. Finally, it moves from 4 to 2 as a free particle with wave function u_2 and momentum $\not{p}_2 = \not{p}_1 + \not{q}_1 + \not{q}_2$. It is also clear from the diagram that the integral need be taken over q_1 only, because when $\not{p}_1$ and $\not{p}_2$ are given, $\not{q}_2$ is determined by $\not{q}_2 = \not{p}_2 - \not{p}_1 - \not{q}_1$. The law of conservation of energy requires $p_1{}^2 = m^2$, $p_2{}^2 = m^2$; but, since the intermediate state is a virtual state, it is not necessary that $(\not{p}_1 + \not{q}_1)^2 = m^2$. Since the operator $1/(\not{p}_1 + \not{q}_1 - m)$ may be resolved as $(\not{p}_1 + \not{q}_1 + m)/[(\not{p}_1 + \not{q}_1)^2 - m^2]$, the importance of a virtual state is inversely proportional to the degree to which the conservation law is violated.

The results given in Eqs. (18-3′) and (18-6) may be summarized by the following list of handy rules† for computing the matrix element $M = (\tilde{u}_2 N u_1)$:

1. An electron in a virtual state of momentum $\not{p}$ contributes the amplitude $i/(\not{p} - m)$ to N.

2. A potential containing the momentum q contributes the amplitude $-ie\not{a}(q)$ to N.

3. All indeterminate momenta q_i are summed over $d^4q_i/(2\pi)^4$.

Remember, in computing the integral, the value of the integral is desired, with the path of integration passing the singularities in a definite manner. Thus replace m by $m - i\epsilon$ in the integrand; then in the solution take the limit as $\epsilon \to 0$.

For relativistic work, only a few terms in the perturbation series are necessary for computation. To assume that fast electrons (and positrons) interact with a potential only once (Born approximation) is often sufficiently accurate.

After the matrix element is determined, the probability of transition per second is given by

$$P = 2\pi/(\Pi N)|M|^2 \times \text{(density of final states)}$$

where ΠN is the normalization factor defined in Lecture 16.

---

†See Summary of numerical factors for transition probabilities, R. P. Feynman, An Operator Calculus, Phys. Rev., **84**, 123 (1951); included in this volume.

Relativistic Treatment of the Interaction of Particles with Light

Nineteenth Lecture

In Lecture 2 the rules governing nonrelativistic interaction of particles with light were given. The rules stated what potentials were to be used in the calculation of transition probabilities by perturbation theory. Those potentials are also applicable to the relativistic theory if the matrix elements are computed as described in Lecture 18. For absorption of a photon, the potential used in nonrelativistic theory was

$$A_\mu = (4\pi e^2)^{1/2} (2\omega)^{-1/2} e_\mu \exp(ik \cdot x) \qquad \begin{cases} K_4 = \omega \\ K \cdot K = 0 \\ \hbar = c = 1 \end{cases} \tag{19-1}$$

For emission of a photon, the complex conjugate of this expression is used. These potentials are normalized to one photon per cubic centimeter and hence the normalization is not invariant under Lorentz transformations. In a manner similar to that for the normalization of electron wave functions, photon potentials will, in the future, be normalized to 2ω photons per cubic centimeter by dropping the $(2\omega)^{-1/2}$ factor in Eq. (19-1), giving

$$A_\mu = (4\pi e^2)^{1/2} e_\mu \exp(ik \cdot x) \tag{19-1'}$$

This makes any matrix element computed with these potentials invariant, but to obtain the correct transition probability in a given coordinate system, it is necessary to reinsert a factor $(2\omega)^{-1}$ for each photon in the initial and final states. This becomes part of the normalization factor ΠN, which contains a similar factor for each electron in the initial and final states.

In momentum representation, the amplitude to absorb (emit) a photon of polarization e_μ is $-i(4\pi e^2)^{1/2}\not{e}$. The polarization vector e_μ is a unit vector perpendicular to the propagation vector. Hence $e \cdot e = -1$ and $e \cdot q = 0$.

RADIATION FROM ATOMS

The transition probability per second is

$$\text{Trans. prob./sec} = 2\pi\,|H|^2 \times (\text{density of final states})$$

where H is the matrix element of the relativistic Hamiltonian,

$$H = \alpha \cdot (-i\nabla - e\mathbf{A}) \qquad \text{S.R.}$$

between initial and final states. That is,

$$\langle f|H|i\rangle = (4\pi e^2)^{1/2} \int \Psi_f{}^* [\alpha \cdot e \exp(i\mathbf{k} \cdot \mathbf{x})]\, \Psi_i \; d\,\text{vol} \qquad (19\text{-}2)$$

Problem: Show that in the nonrelativistic limit, Eq. (19-2) reduces to

$$1/2m \int \Psi_f{}^* [e \cdot p \exp(i\mathbf{k} \cdot \mathbf{x}) + \exp(i\mathbf{k} \cdot \mathbf{x})\, p \cdot e + e \cdot (\sigma \times \mathbf{k})$$
$$\times \exp(i\mathbf{k} \cdot \mathbf{x})]\; \Psi_i \; d\,\text{vol}$$

This is the same result as was obtained from the Pauli equation.

SCATTERING OF GAMMA RAYS BY ATOMIC ELECTRONS

A relativistic treatment of scattering of photons from electrons will now be given. As an approximation, consider the electrons to be free (energies at which a relativistic treatment is necessary are, generally, much greater than atomic binding energies). This will lead to the Klein-Nishina formula for the Compton-effect cross section.

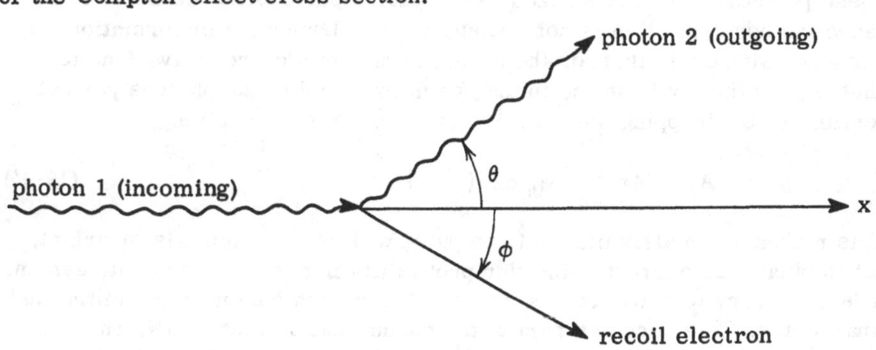

FIG. 19-1

For the incoming photon take as a potential $A_{1\mu} = e_{1\mu} \exp(-iq_1 \cdot x)$ and for the outgoing photon take $A_{2\mu} = e_{2\mu} \exp(-iq_2 \cdot x)$. The light is polarized perpendicular to the direction of propagation (see Fig. 19-1). Thus,

$$e_1 \cdot q_1 = 0 \qquad e_2 \cdot q_2 = 0$$

also

$$q_1 \cdot q_1 = q_1^2 = 0 \qquad \text{and} \qquad q_2 \cdot q_2 = q_2^2 = 0 \tag{19-3}$$

As initial and final state electron wave functions, choose

$$\Psi_1 = u_1 \exp(-ip_1 \cdot x)$$

$$\Psi_2 = u_2 \exp(-ip_2 \cdot x)$$

where u_1, u_2, p_1, and p_2 satisfy

$$\not{p}_1 u_1 = m u_1 \qquad \not{p}_2 u_2 = m u_2$$

$$p_1 \cdot p_1 = m^2 \qquad p_2 \cdot p_2 = m^2 \tag{19-4}$$

Conservation of energy and momentum (four equations) is written

$$\not{p}_1 + \not{q}_1 = \not{p}_2 + \not{q}_2 \tag{19-5}$$

If the coordinate system is chosen so that electron number 1 is at rest,

$$\not{p}_1 = m\gamma_t \tag{19-6a}$$

$$\not{p}_2 = E_2\gamma_t - p_2 \cos\phi\gamma_x + p_2 \sin\phi\gamma_y \tag{19-6b}$$

$$\not{q}_1 = \omega_1(\gamma_t - \gamma_x) \tag{19-6c}$$

$$\not{q}_2 = \omega_2(\gamma_t - \gamma_x \cos\theta - \gamma_y \sin\theta) \tag{19-6d}$$

The latter two equations follow from the fact that, for a photon, the energy and momentum are both equal to the frequency (in units in which $c = 1$). The momentum has been resolved into components. The incoming photon beam can be resolved into two types of polarization, which will be designated type A and type B:

$$\text{(A)}\ \not{e}_1 = \gamma_z \qquad \text{(B)}\ \not{e}_1 = \gamma_y$$

Type A has the electric vector in the z direction and type B has the electric vector in the y direction. Similarly the outgoing photon beam can be resolved into two types of polarization:

$$(\text{A}')\ \not{e}_2 = \gamma_z \qquad (\text{B}')\ \not{e}_2 = \gamma_y \cos\theta - \gamma_x \sin\theta$$

Conservation of energy of momentum dictates that either the angle of the recoil electron ϕ or the angle at which the scattered photon comes off θ completely determines the remaining quantities. If the electron direction is unimportant, its momentum can be eliminated by solving Eq. (19-5) for $\not{p}_2$ and squaring the resulting equation:

$$\not{p}_2 = \not{p}_1 + \not{q}_1 - \not{q}_2$$

$$\begin{aligned} p_2^2 = m^2 &= (\not{p}_1 + \not{q}_1 - \not{q}_2)(\not{p}_1 + \not{q}_1 - \not{q}_2) \\ &= p_1^2 + q_1^2 + q_2^2 + 2p_1\cdot q_1 - 2p_1\cdot q_2 - 2q_1\cdot q_2 \\ &= m^2 + 0 + 0 + 2m\omega_1 - 2m\omega_2 - 2\omega_1\omega_2(1 - \cos\theta) \end{aligned}$$

where the last line was obtained from the preceding line by using Eqs. (19-3), (19-4), and (19-6a, c, d). This can be written

$$m(\omega_1 - \omega_2) = \omega_1\omega_2(1 - \cos\theta)$$

or

$$(m/\omega_2) - (m/\omega_1) = 1 - \cos\theta \tag{19-7}$$

This is the well-known formula for the Compton shift in wavelength (or frequency).

DIGRESSION ON THE DENSITY OF FINAL STATES

By the method discussed in the earlier part of the course, the following final state densities (per unit energy interval) can be obtained. When a system of total energy E and total linear momentum **p** disintegrates into a *two-particle final state*,

$$\text{Density of states} = (2\pi)^{-3}E_1E_2\,\frac{p_1^3\,d\Omega_1}{Ep_1^2 - E_1(\mathbf{p}\cdot\mathbf{p}_1)} \tag{D-1}$$

where E_1 = energy of particle 1; E_2 = energy of particle 2; $\mathbf{p}_1$ = momentum of particle 1; $d\Omega_1$ = solid angle, into which particle 1 comes out; m_1 = mass of particle 1; m_2 = mass of particle 2; and $E_1 + E_2 = E$, $\mathbf{p}_1 + \mathbf{p}_2 = \mathbf{p}$.

Another useful formula is in terms of the final energy of particle 1 and its azimuth ϕ_1 (instead of θ_1, ϕ_1). It is

$$\text{Density of states} = (2\pi)^{-3}(E_1E_2/|\mathbf{p}|)\,dE_1\,d\phi_1 \tag{D-2}$$

Special cases: (a) When $m_2 = \infty$ ($E_2 = \infty$, $E = \infty$):

$$\text{Density of states} = (2\pi)^{-3}\, E_1 |p_1|\, d\Omega_1 \tag{D-3}$$

(b) In center-of-mass system $\mathbf{p} = 0$:

$$\text{Density of states} = (2\pi)^{-3}\,[E_1 E_2\, d\Omega_1/(E_1 + E_2)] \tag{D-4}$$

When a system disintegrates into a three-particle final state,

$$\text{Density of states} = (2\pi)^{-6}\, E_3 E_2 \frac{p_2^3 p_1^2\, dp_1\, d\Omega_1\, d\Omega_2}{p_2^2 (E - E_1) - E_2 \mathbf{p}_2 \cdot (\mathbf{p} - \mathbf{p}_1)} \tag{D-5}$$

Special case: When $m_3 = \infty$:

$$\text{Density of states} = (2\pi)^{-6} E_2 |p_2|\, d\Omega_2\, p_1^2\, dp_1\, d\Omega_1 \tag{D-6}$$

The Compton effect has a two-particle final state: taking particle 1 to be photon 2 and particle 2 to be electron 2, from Eq. (D-1),

$$\text{Density of states} = (2\pi)^{-3}\, \omega_2 E_2 \frac{\omega_2^3\, d\Omega_\omega}{(m + \omega_1)\omega_2^2 - \omega_2(\omega_1\omega_2 \cos\theta)}$$

COMPTON RADIATION

Calculation of $|M|^2$. Using the Compton relation Eq. (19-7) to eliminate θ, this becomes

$$\text{Density of states} = (2\pi)^{-3}\,(E_2\omega_2^3\, d\Omega_\omega/m\omega_1)$$

The probability of transition per second is given by

$$\text{Trans. prob./sec} = \sigma c = (2\pi/2E_1 2E_2 2\omega_1 2\omega_2)\,|M|^2$$

$$\times\, (2\pi)^{-3}(E_2\omega_2^3\, d\Omega_\omega/m\omega_1)$$

or

$$\sigma = [\omega_2^2\, d\Omega_\omega/(2\pi)^2 16 m^2 \omega_1^2]\,|M|^2$$

In working out the matrix element M, there are two ways in which the scattering can happen: (R) the incoming photon is absorbed by the electron and then the electron emits the outgoing photon; (S) the electron emits a photon and subsequently absorbs the incident photon. These two processes are shown diagrammatically in Fig. 19-2.

In momentum representation, the matrix element M for the first process R is

$$i[-i(4\pi e^2)^{1/2}]^2 \{\tilde{u}_2 \not{e}_2 [1/(\not{p}_1 + \not{q}_1 - m)] \not{e}_1 u_1\}$$

Reading from right to left the factors in the matrix element are interpreted as follows: (a) The initial electron enters with amplitude u_1; (b) the electron is first scattered by a potential (i.e., absorbs a photon); (c) having re-

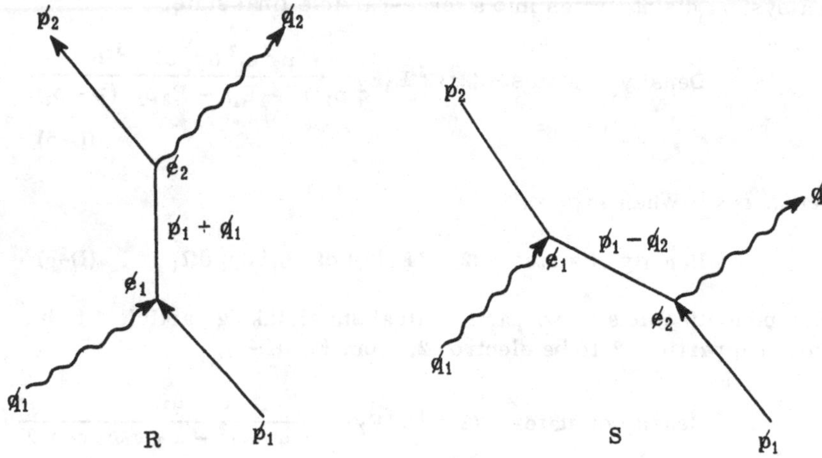

FIG. 19-2

ceived momentum $\not{q}_1$ from the potential the electron travels as a free electron with momentum $\not{p}_1 + \not{q}_1$; (d) the electron emits a photon of polarization $\not{e}_2$; and (e) we now ask for the amplitude, that the electron is in a state u_2.

Exercise: Write down the matrix element for the second process S. The total matrix element is the sum of these two. Rationalize these matrix elements and, using the table of matrix elements (Table 13-1) work out $|M|^2$.

Twentieth Lecture

For the R diagram, M was found to be

$$-i4\pi e^2 \{\tilde{u}_2 \not{e}_2 [1/(\not{p}_1 + \not{q}_1 - m)] \not{e}_1 u_1\} = -i4\pi e^2 (\tilde{u}_2 R u_1)$$

and as an exercise the matrix element for the S diagram was found to be

$$-i4\pi e^2\{\tilde{u}_2\not{e}_1[1/(\not{p}_1 - \not{q}_2 - m)]\not{e}_2 u_1\} = -i4\pi e^2(\tilde{u}_2 S u_1)$$

The complete matrix element is the sum of these, so that the cross section becomes

$$\sigma = (e^4/4m)(\omega_2{}^2/\omega_1{}^2)\, d\Omega_2\, |\tilde{u}_2(R+S)u_1|^2$$

The problem now is actually to compute the matrix elements for R and S. First R will be considered. Using the identity

$$1/(\not{p} - m) = (\not{p} + m)/(p^2 - m^2)$$

the matrices may be removed from the denominator of R giving

$$R = \frac{\not{e}_2(\not{p}_1 + \not{q}_1 + m)\not{e}_1}{(\not{p}_1 + \not{q}_1) \;-\; m} = \frac{\not{e}_2(\not{p}_1 + \not{q}_1 + m)\not{e}_1}{2m\omega_1}$$

The denominator is seen to be $2m\omega_1$ from the following relations:

$$(\not{p}_1 + \not{q}_1)^2 - m^2 = p_1{}^2 + 2p_1 \cdot q_1 + q_1{}^2 - m^2$$

$$p_1{}^2 = m^2$$

$$q_1{}^2 = 0$$

$$2p_1 \cdot q_1 = 2m\omega_1$$

The matrix elements for the various spin and polarization combinations can be calculated straightforwardly from this point. But certain preliminary manipulations will reduce the labor involved. Using the identity

$$\not{a}\not{b} = 2a \cdot b - \not{b}\not{a}$$

it is seen that

$$\not{e}_2\not{p}_1\not{e}_1 = \not{e}_2(2p_1 \cdot e_1) - \not{e}_2\not{e}_1\not{p}_1$$

But p_1 has only a time component and e_1 only a space component so $p_1 \cdot e_1 = 0$. Recalling that $\not{p}_1 u_1 = m u_1$, it is seen that

$$\tilde{u}_2\not{e}_2\not{p}_1\not{e}_1 u_1 = -\tilde{u}_2\not{e}_2\not{e}_1\not{p}_1 u_1 = -(\tilde{u}_2\not{e}_2\not{e}_1 u_1)m$$

and this is the matrix element of the first term of R. It is also the negative of the matrix element of the last term of R, so R may be replaced by the equivalent

$$R = \not{e}_2\not{q}_1\not{e}_1/2m\omega_1$$

By an exactly similar manipulation, the S matrix is equivalent to

$$S = \not{e}_1 \not{q}_2 \not{e}_2 / 2m\omega_2$$

Substituting $\not{q}_1 = \omega_1(\gamma_t - \gamma_x)$ and $\not{q}_2 = \omega_2(\gamma_t - \gamma_x \cos\theta - \gamma_y \sin\theta)$ and transposing the 2m factor, the complete matrix may be written

$$2m(R+S) = \not{e}_2(\gamma_t - \gamma_x)\not{e}_1 + \not{e}_1(\gamma_t - \gamma_x \cos\theta - \gamma_y \sin\theta)\not{e}_2$$

A still more useful form is obtained by noting that $\not{e}_1$ anticommutes with q_1 ($e_1 \cdot q_1 = 0$) and $\not{e}_2$ with q_2 and that $\not{e}_2\not{e}_1 = 2e_2 \cdot e_1 - \not{e}_1\not{e}_2$. Thus,

$$2m(R+S) = -\not{e}_2\not{e}_1(\gamma_t - \gamma_x) - \not{e}_1\not{e}_2(\gamma_t - \gamma_x + \gamma_x - \gamma_x \cos\theta - \gamma_y \sin\theta)$$

$$= -2(e_2 \cdot e_1)(\gamma_t - \gamma_x) - \not{e}_1\not{e}_2[\gamma_x(1 - \cos\theta) - \gamma_y \sin\theta]$$

Using this form of the matrix, the matrix elements may be computed easily. For example, consider the case for polarization: $\not{e}_1 = \gamma_z$, $\not{e}_2 = \gamma_y \cos\theta - \gamma_x \sin\theta$. This corresponds to cases (A) and (B′) of Lecture 19 and will be denoted by (AB′). The matrix is

$$2m(R+S) = -\gamma_z(\gamma_y \cos\theta - \gamma_x \sin\theta)[\gamma_x(1 - \cos\theta) - \gamma_y \sin\theta]$$

since $e_2 \cdot e_1 = 0$. Expanded this becomes

$$2m(R+S) = -\gamma_z[\gamma_y\gamma_x \cos\theta(1 \cos\theta) + \cos\theta \sin\theta + \sin\theta(1 - \cos\theta)$$

$$+ \gamma_x\gamma_y \sin^2\theta]$$

$$= -\gamma_z(\gamma_x\gamma_y - \gamma_x\gamma_y \cos\theta + \sin\theta) = -\gamma_x\gamma_y\gamma_z(1 - \cos\theta)$$

$$- \gamma_z \sin\theta$$

where the anticommutation of the γ's has been used. In the case of spin-up for the incoming particle and spin down for the outgoing particle ($s_1 = -1$), $s_2 = -1$), the matrix elements

$$-2m(F_1F_2)^{1/2}(\tilde{u}_2\gamma_x\gamma_y\gamma_z u_1) = -iF_2 p_{1+} - iF_1 p_{2+}$$

$$-2m(F_1F_2)^{1/2}(\tilde{u}_2\gamma_z u_1) = +p_{1+}F_2 - p_{2+}F_1$$

may be found by reference to Table 13-1. But note that in this problem $p_{1+} = p_{x1} + ip_{y1} = 0$ since particle 1 is at rest. Hence the final matrix element for this case, polarization (AB′), spin $s_1 = +1$, $s_2 = -1$, is

TABLE 20-1

Polarization		AA′	AB′	BA′	BB′
$\not{e}_1$		γ_z	γ_z	γ_y	γ_y
$\not{e}_2$		γ_z	$\gamma_y \cos\theta - \gamma_x \sin\theta$	γ_z	$\gamma_y \cos\theta - \gamma_x \sin\theta$
Matrix $2m(R + S)$		$2\gamma_t - \gamma_x(1 + \cos\theta) - \gamma_y \sin\theta$	$-\gamma_x\gamma_y\gamma_z(1 - \cos\theta) - \gamma_z \sin\theta$	$-\gamma_x\gamma_y\gamma_z(1 - \cos\theta) + \gamma_z \sin\theta$	$2\cos\theta\,\gamma_t - \gamma_x(1 + \cos\theta) - \gamma_y \sin\theta$
Matrix elements $2m\sqrt{F_1F_2}\,(\tilde{u}_2(R+S)u_1)$	$s_1 = +1$	$+2F_2F_1 - (1 + \cos\theta)F_1p_{2-}$	0	0	$2\cos F_2F_1 - (1 + \cos\theta)F_1p_{2-}$
	$s_2 = +1$	$-i \sin\theta\ F_1p_{2-}$	0	0	$-i \sin\theta\ F_1p_{2-}$
	$s_1 = +1$	0	$-i(1 - \cos\theta)\ F_1p_{2+}$	$-i(1 - \cos\theta)F_1p_{2+}$	0
	$s_2 = -1$	0	$-\sin\theta\ F_1p_{2+}$	$+\sin\theta\ F_1p_{2+}$	0

Note: The matrix elements for $\begin{pmatrix} s_1 = -1 \\ s_2 = -1 \end{pmatrix}$ are the complex conjugates of those above for $\begin{pmatrix} s_1 = +1 \\ s_2 = +1 \end{pmatrix}$, and for $\begin{pmatrix} s_1 = -1 \\ s_2 = +1 \end{pmatrix}$ they are the complex conjugates of those for $\begin{pmatrix} s_1 = +1 \\ s_2 = -1 \end{pmatrix}$ above.

$$2m\,(F_1F_2)^{1/2}\,(\tilde{u}_2(R+S)u_1) = -(1-\cos\theta)iF_1p_{2+} - \sin\theta\ p_{2+}F_1$$

The results for the other combinations of polarization and spin are obtained in the same manner and will only be presented in tabular form (Table 20-1). They may be verified as an exercise.

For any one of the polarization cases listed, $|M|^2$ is the sum of the square amplitudes of the matrix elements for outgoing spin states averaged over incoming spin states. But this is seen to be simply the square magnitude of the nonzero matrix element listed under the appropriate polarization case. For example, in case (AA′),

$$|M|^2 = |\tilde{u}_2(R+S)u_1|^2 = (1/4m^2F_1F_2)\,|2F_2F_1 - (1+\cos\theta)F_1p_{2-} - i\sin\theta\ F_1p_{2+}$$

By employing the relation

$$p_{2-} = p_{1-} + q_{1-} - q_{2-} = q_{1-} - q_{2-} = \omega_1 - \omega_2\cos\theta + i\omega_2\sin\theta$$

and

$$(m/\omega_2) - (m/\omega_1) = 1 - \cos\theta$$

the square magnitudes of the matrix elements for the various cases reduce, after considerable amount of algebra, to the expressions given in Table 20-2.

TABLE 20-2

Polarization	$\|M\|^2$
AA′	$[(\omega_1-\omega_2)^2/\omega_1\omega_2] + 4$
AB′	$[(\omega_1-\omega_2)^2/\omega_1\omega_2]$
BA′	$[(\omega_1-\omega_2)^2/\omega_1\omega_2]$
BB′	$[(\omega_1-\omega_2)^2/\omega_1\omega_2] + 4\cos^2\theta$

It is clear that all four of these formulas may be written simultaneously in the form

$$|M|^2 = [(\omega_1-\omega_2)^2/\omega_1\omega_2] + 4(e_1\cdot e_2)^2$$

Note that these formulas are not adequate for circular polarization. That is, if $\not{e}_1$ were, for example, $1/\sqrt{2}\,(i\gamma_z + \gamma_y)$, it is seen that because of the phasing represented by the imaginary part of $\not{e}_1$, all the calculations must be carried out *before* squaring the matrix elements in order to get the proper interference.

Finally the cross section for scattering with prescribed plane polarization of the incoming and outgoing photons is

$$\sigma = (e^4/4m^2)(\omega_2{}^2/\omega_1{}^2)\ d\Omega\,\omega_2\ [(\omega_2/\omega_1) + (\omega_1/\omega_2) - 2 + 4(e_1 \cdot e_2)^2]$$

This is the Klein-Nishina formula for polarized light. For unpolarized light this cross section must be averaged over all polarizations.

It is noted that diagram cases such as Fig. 20-1 have been included in

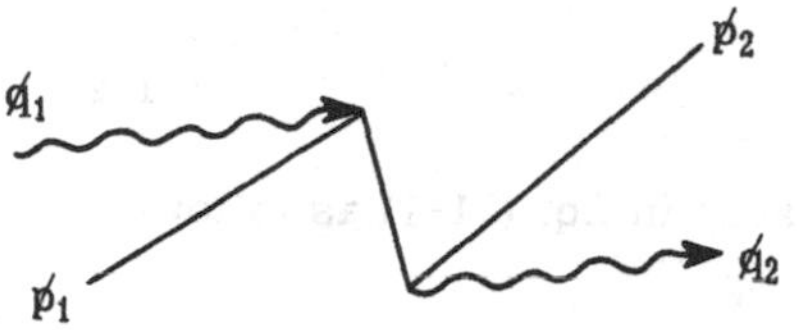

FIG. 20-1

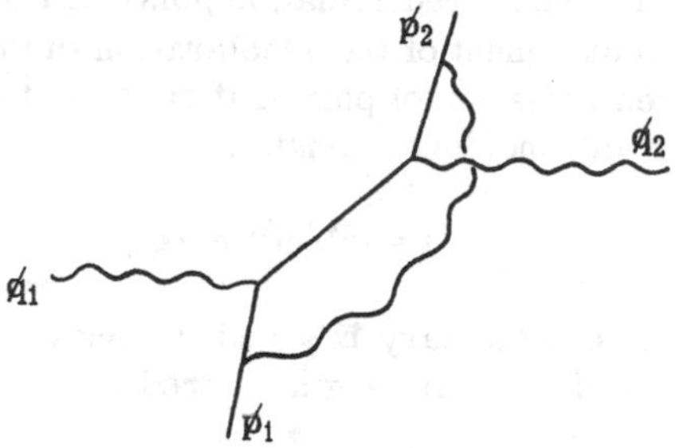

FIG. 20-2

the previous derivation as a result of the generality in the transformation of of $K_+(2,1)$ to momentum representation. In fact, all diagram cases have been included except higher-order effects to be discussed later. (They correspond to emission and reabsorption of a third photon by the electron, such as in Fig. 20-2.)

Twenty-first Lecture

Discussion of the Klein-Nishina Formula. In the "Thompson limit," $\omega_1 << m$. Tnen the electron picks up very little energy in recoil, and $\omega_1 \approx \omega_2$. This can be seen from the relation

$$m\omega_1 - m\omega_2 = \omega_1\omega_2(1 - \cos\theta) \tag{21-1}$$

In this limit, the Klein-Nishina formula gives

$$\sigma = (e^4/m^2)(e_1 \cdot e_2)^2\ d\Omega_\omega \tag{21-2}$$

which is the Rayleigh-Thompson scattering cross section. Note that ω is still very large compared to the eigenvalues of an atom, in accordance with our original assumptions for Compton scattering.

The same result is obtained by a classical picture. Under the action of the electric field of the photon $\mathbf{E} = E_0 \mathbf{e}_1 \exp(i\omega t)$, the electron is given the acceleration

$$\mathbf{a} = (e/m)E_0 \mathbf{e}_1 \exp(i\omega t)$$

Classically, an accelerated charge radiates to give the scattered radiation

$$\mathbf{E}_s = -\frac{e}{R} \text{ (retarded acceleration projected on plane } \perp \text{ to line of sight)}$$

The scattered radiation polarized in the direction $\mathbf{e}_2$ is determined by the component of the acceleration in this direction. The intensity of the scattered radiation of polarization $\mathbf{e}_2$ is then (times R^2 per unit solid angle and per unit incident intensity)

$$I = (e^4/m^2)(\mathbf{e}_1 \cdot \mathbf{e}_2)^2 \tag{21-2'}$$

The customary $\hbar$'s and c's may be replaced in Eq. (21-1) as follows (σ is an area or length squared):

$$e^4 = (e^2)^2 = (e^2/\hbar c)^2$$

$$m^2 = (mc/\hbar)^2 = \text{length squared}$$

$$e^4/m^2 = (e^2/mc^2)^2 = r_0^2 \approx 8 \times 10^{-26} \text{ cm}^2$$

Averaging over Polarizations. It is often desired to have the scattering cross section for a beam regardless of the incoming or outgoing polarization. This can be obtained by summing the probabilities over the polarizations of the outgoing beam and averaging over the incoming beam. Thus, suppose the incoming beam has polarization of type A. The probabilities (or cross sections) for the two possible types of outgoing polarization, A′ and B′ can be symbolized as AA′ and AB′. The total probability for scattering a photon of either polarization is AA′ + AB′. Then suppose the incoming beam is equally likely to be polarized as type A or type B. The resulting probability can be obtained as the sum 1/2 (probability if type A) + 1/2 (probability if type B). This is the situation for unpolarized incoming beam, and gives

$$\begin{aligned}\sigma\,(\text{averaged over polarizations}) &= (1/2)(AA' + AB') + (1/2)(BA' + BB') \\ &= \frac{e^4}{2m^2}\left(\frac{\omega_2}{\omega_1}\right)^2 d\Omega_{\omega_2}\left(\frac{\omega_2}{\omega_1} + \frac{\omega_1}{\omega_2} - \sin^2\theta\right)\end{aligned} \tag{21-3}$$

If, on the other hand, the polarization of the outgoing beam is measured (still with an unpolarized incoming beam), its dependence on frequency and scattering angle is given by the ratio

$$\frac{\text{Probability of polarization type A}'}{\text{Probability of polarization type B}'} = \frac{(1/2)[AA' + BA']}{(1/2)[AB' + BB']}$$

$$= \frac{(\omega_2/\omega_1) + (\omega_1/\omega_2)}{(\omega_2/\omega_1) + (\omega_1/\omega_2) - 2\sin^2\theta}$$

The forward radiation ($\theta = 0$) remains unpolarized, but a certain degree of polarization will be found in light scattered through any nonzero angle. In the low-frequency limit ($\omega_1 \approx \omega_2$), the polarization is complete at $\theta = \pi/2$. Thus an unpolarized beam becomes plane-polarized when scattered through 90°.†

Total Scattering Cross Section. If the cross section (averaged over polarizations) given in Eq. (21-3) is integrated over the solid angle

$$d\Omega = 2\pi\, d(\cos\theta) = (2\pi m/\omega_2^2)\, d\omega_2$$

the total cross section for scattering through any angle is obtained. So, from Eq. (21-1),

$$\cos\theta = 1 - m/\omega_2 + m/\omega_1 \tag{21-1$'$}$$

and the variable ω_2 goes between the limits $m\omega_1/(2\omega_1 + m)$ and ω_1 as $\cos\theta$ goes from -1 to $+1$. Equation (21-3) can be written

$$d\sigma_T = (e^4/2m^2)(2\pi/\omega_1^2)m\, d\omega_2(\omega_2/\omega_1 + \omega_1/\omega_2 - 2m/\omega_2 + 2m/\omega_1$$
$$+ m^2/\omega_1^2 + m^2/\omega_2^2 - 2m^2/\omega_1\omega_2)$$

where the last five terms replace $-\sin^2\theta = \cos^2\theta - 1$ using Eq. (21-1′). Simple integrations yield‡

$$\sigma_T = \pi e^4/m^2[(m/\omega_1 - 2m^2/\omega_1^2 - 2m^3/\omega_1^3)\log(2\omega_1/m + 1)$$
$$+ m/2\omega_1 + 4m^2/\omega_1^2 - m^3/2\omega_1(2\omega_1 + m)^2]$$

In the high-frequency limit ($\omega_1 \rightarrow \infty$)

$$\sigma_T \sim (1/\omega_1)\log\omega_1 \rightarrow 0$$

---

† Cf. Walter Heitler, "Quantum Theory of Radiation," 3rd ed., Oxford, 1954; and B. Rossi and K. Greissen, Phys. Rev., 61, 121 (1942).

‡ Cf. Heitler, op. cit., p. 53.

Thus Compton scattering is a negligible effect at high frequencies, where pair production becomes the important effect.

TWO-PHOTON PAIR ANNIHILATION

From the quantum-electrodynamical point of view, another phenomenon completely analogous to Compton scattering is two-photon pair annihilation. Two photons are necessary (in the outgoing radiation) to maintain conservation of momentum and energy when pair annihilation takes place in the absence of an external potential. The interaction can be diagrammed as shown in Fig. 21-1. This figure should be compared to that for Compton scattering (Lecture 20). The only differences are that the direction of photon $\not{q}_1$ is reversed, and, since particle 2 is a positron, $\not{p}_2 = -$(momentum of positron). So write

$$\not{p}_1 = (E_-\gamma_t - \mathbf{p}_- \cdot \boldsymbol{\gamma})$$

$$\not{p}_2 = -(E_+\gamma_t - \mathbf{p}_+ \cdot \boldsymbol{\gamma})$$

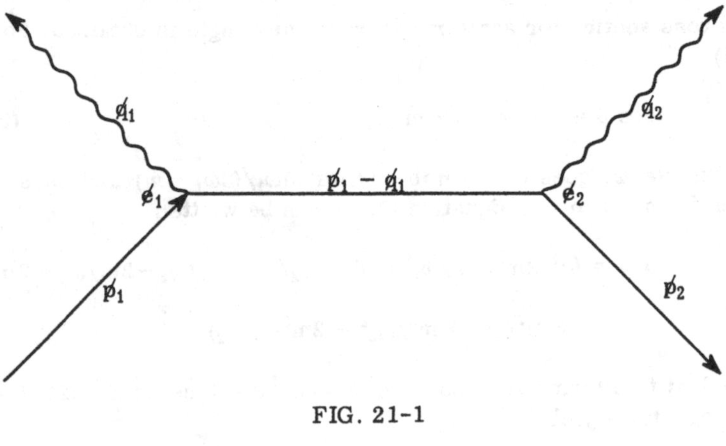

FIG. 21-1

where the energies E_- and E_+ of the electron and positron are both positive numbers. The conservation law gives

$$\not{p}_2 = \not{p}_1 - \not{q}_1 - \not{q}_2 \tag{21-4}$$

(just as for Compton scattering, but the direction of $\not{q}_1$ reversed), so the matrix element for this interaction is

$$M_1 = -i\,4\pi e^2 (\tilde{u}_2 \not{e}_2 (\not{p}_1 - \not{q}_1 - m)^{-1} \not{e}_1 u_1)$$

The second possibility, indistinguishable from the first by any measurement, is obtained from the first by interchanging the two photons (see Fig. 21-2); again note similarity to Compton scattering.

Immediately, the matrix element is

$$M_2 = -i\,4\pi e^2 (\tilde{u}_2 \not{e}_1 (\not{p}_1 - \not{q}_2 - m)^{-1} \not{e}_2 u_1)$$

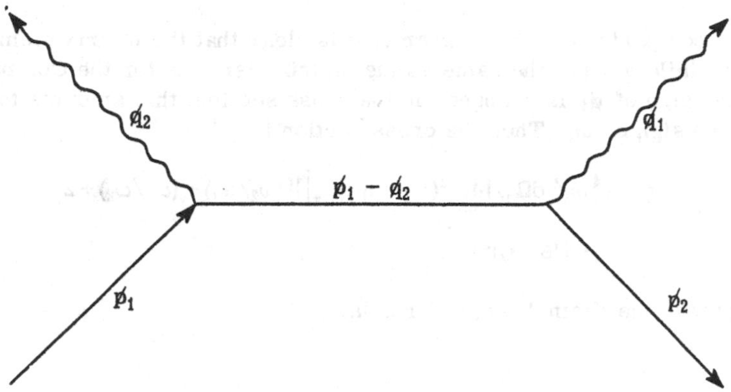

FIG. 21-2

The sum of the two matrix elements and the density of final states gives the cross section

$$\sigma \cdot (\text{velocity of positron}) = 2\pi/(2E_- \cdot 2E_+ \cdot 2\omega_1 \cdot 2\omega_2) \cdot |M_1 + M_2|^2 \times (\text{density of states})$$

in a system where the electron is at rest and the positron is moving. The density of final states is

$$\omega_1 \omega_2/(2\pi)^3\, \omega_1^{\,2}\, d\Omega_1/(\omega_2\omega_1 - Q_2 \cdot Q_1)$$

Since particle 2 is a positron, $\not{p}_2 = -\not{p}_+$, so the conservation law, Eq. (21-4), gives

$$\not{p}_1 + \not{p}_+ = \not{q}_1 + \not{q}_2$$

Then

$$m^2 + 2(p_1 \cdot p_+) + m^2 = 0 + 2q_1 \cdot q_2 + 0$$

This reduces to

$$2m^2 + 2mE_+ = 2\omega_1\omega_2 - 2Q_1 \cdot Q_2$$

Taking the velocity of the positron as $|\mathbf{p}_+|/E_+$, the cross section is

$$\sigma = (2\pi)\,\omega_1^2\, d\Omega_1/[2E_- \cdot 2\,|\mathbf{p}_+|\, 4(2\pi)^3 \cdot m(E_+ + m)] \times |M_1 + M_2|^2$$

$$= \frac{\omega_1^2\, d\Omega_1\, |M_1 + M_2|^2}{64\pi^2\; m^2\, |\mathbf{p}_+|\, (m + E_+)}$$

From a comparison of the diagrams, it is clear that the matrix elements for pair annihilation are the same as the matrix elements for the Compton effect if the sign of $\not{q}_1$ is changed. In the cross section, this amounts to changing the sign of ω_1. Then the cross section is

$$\sigma = e^4\,\omega_1^2\, d\Omega_1/[4m^2(E_+ + m)|\mathbf{p}_+|][(\omega_2/\omega_1) + (\omega_1/\omega_2) + 2$$
$$- 4(\mathbf{e}\cdot\mathbf{e}_2)^2]$$

in analogy with the Klein-Nishina formula.

Twenty-second Lecture

POSITRON ANNIHILATION FROM REST

The formula for positron-electron annihilation derived in Lecture 21 diverges as the positron velocity approaches zero ($\sigma \sim 1/v$; this is true for other cross sections when a process involves absorption of the incoming particle, and is the well-known $1/v$ law). To calculate the positron lifetime in an electron density ρ (recall that the preceding cross section was for an electron density of one per cubic centimeter) as $v_+ \to 0$, we use

$$\text{Trans. prob./sec} = \sigma v_+ \rho$$

plus the fact that, as $v_+ \to 0$, $E_+ \to m$ and $\omega_1 \to \omega_2 \to m$ (when the electron and positron are both approximately at rest, momentum and energy can be conserved only with two photons of momenta equal in magnitude but opposite in direction). Thus

$$\text{Trans. prob./sec} = \sigma v_+ \rho = (e^4/2m^2)\,\rho\, d\Omega\,(\sin^2\theta) \tag{22-1}$$

where θ = angle between directions of polarization of two photons ($\cos\theta = \mathbf{e}_1 \cdot \mathbf{e}_2$). The $\sin^2\theta$ dependence indicates that the two photons have their polarizations at right angles. To get the probability of transition per second for any photon direction and any polarization, it is necessary to sum over solid angle ($\int d\Omega = 4\pi$) and average over polarizations ($\sin^2\theta = 1/2$), giving

$$\text{Total trans. prob./sec} = 1/\tau = (\pi e^4/m^2)\,\rho$$
$$= \pi(e^2/mc^2)^2 c\rho = \pi r_0^2 c\rho \tag{22-2}$$

(factors of c and ħ reinserted where required), where r_0 = classical electron radius, and τ = mean lifetime.

Problems: (1) Obtain the preceding result directly by using matrix elements for an electron and positron at rest. Show that only the singlet state (spins antiparallel) can disintegrate into two photons. The triplet state disintegrates into three photons and has a longer lifetime (see the next problem).

(2) Find the mean time required for a positron and electron to disintegrate into three photons (spins must be parallel). The following procedure is suggested: (1) set up formula for rate of disintegration; (2) write M in the simplest possible form; (3) make a table of matrix elements (same as Table 13-1 but with $\not{p}_1 = m\gamma_t$, $\not{p}_2 = -m\gamma_t$); (4) find the matrix element of M for eight polarization cases; (5) find the rate of disintegration for each case; (6) sum the disintegration rate over polarizations; (7) obtain the photon spectrum; (8) obtain the total disintegration rate by integrating over photon spectrum and angle; and (9) compare with Orr and Powel.†

(3) It is known that the matrix elements should be independent of a gauge transformation $\not{e}' = \not{e} + \alpha \not{q}$, where α is an arbitrary constant and $\not{q}$ is the momentum of a photon whose polarization is $\not{e}$ or $\not{e}'$. Show that substituting $\not{q}$ for $\not{e}$ in the matrix elements for the Compton effect gives m = 0.

BREMSSTRAHLUNG

When an electron passes through the Coulomb field of a nucleus it is deflected. Associated with this deflection is an acceleration which, according to the classical theory, results in radiation. According to quantum electrodynamics, there is a certain probability that the incident electron will make a transition to a different electron state with a photon emitted, while in the field of the nucleus. Interaction with the field of the nucleus is necessary to satisfy conservation of energy and momentum. That is, the electron cannot emit a photon and make a transition to a different electron state while traveling along in a vacuum. Figure 22-1 shows the process and defines angles that arise later.

The Coulomb potential of the nucleus will be considered to act only once (Born approximation). The validity of this approximation was discussed in Lecture 16. There are two (indistinguishable) orders in which the bremsstrahlung process can occur: (a) the electron interacts with the Coulomb field and subsequently emits a photon, or (b) the electron first emits a photon and then interacts with the Coulomb field. The diagrams for these proc-

† A. Ore and J. L. Powell, Phys. Rev., 75, 1696 (1949).

esses are shown in Fig. 22-2. The interaction with the nucleus gives momentum $Q\!\!\!/$ to the electron. Conservation of energy and momentum requires

$$p\!\!\!/_1 + Q\!\!\!/ = p\!\!\!/_2 + q\!\!\!/ \qquad \text{or} \qquad Q\!\!\!/ = p\!\!\!/_2 - p\!\!\!/_1 + q\!\!\!/$$

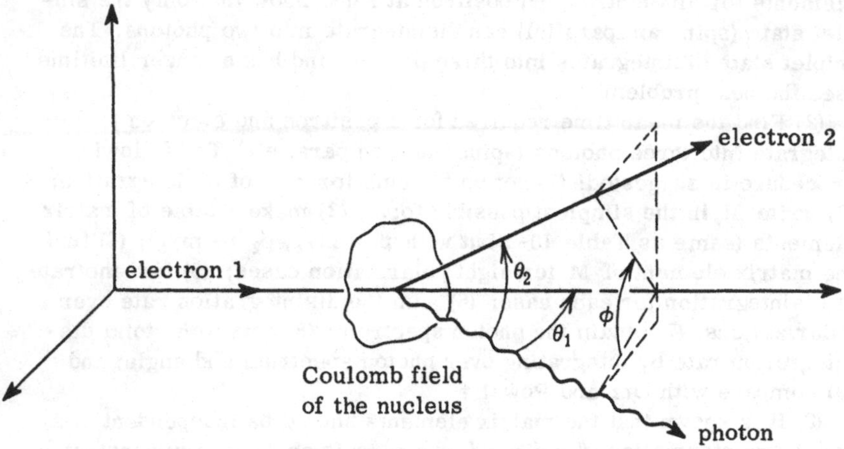

FIG. 22-1

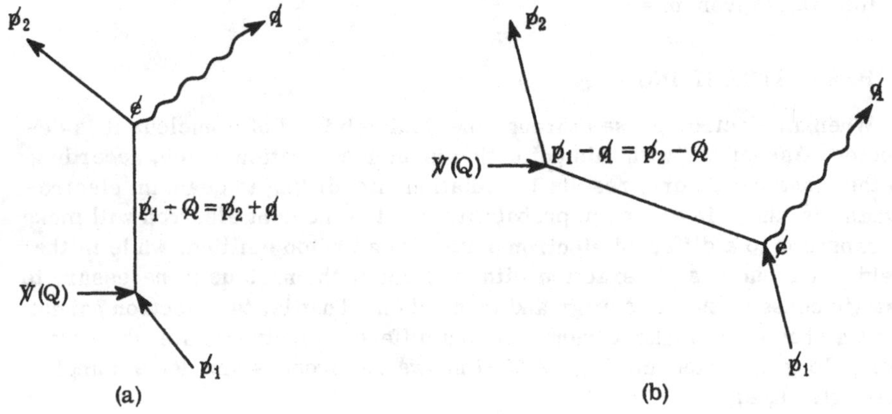

FIG. 22-2

In Lecture 18 it was shown that the Fourier transform of the Coulomb potential was proportional to $\delta(Q_4)$, since the potential is independent of time. This means that only transitions for which $Q_4 = 0$ occur, or energy must be conserved among the incident electron, final electron, and photon. Thus $E_1 = E_2 + \omega$. The transition probability is given by

$$\text{Trans. prob./sec} = \sigma v_1 = (2\pi/2E_1 2E_2 2\omega)\, |\mathfrak{M}|^2 \times D$$

Since the nucleus is to be considered infinitely heavy,

$$D = (2\pi)^{-6}\, E_2 p_2\, d\Omega_2 \omega^2\, d\omega\, d\Omega_\omega$$

Notice that there is a spectrum of photons; that is, the photon energy is not determined (as it was in the Compton effect, for example). Letting $\mathfrak{M} = (\tilde{u}_2 M u_1)$,

$$M = (-i)(4\pi e^2)^{1/2} \left[\not{e} \frac{1}{p_1 + Q - m} \not{V}(Q) + \not{V}(Q) \frac{1}{p_2 - Q - m} \not{e} \right] \tag{22-3}$$

where the first term comes from Fig. 22-2a and the second term from Fig. 22-2b. The explanation of the factors in the first term, for example, is, reading from right to left, that an electron initially in state u_1 is scattered by the Coulomb potential acquiring an additional momentum $\not{Q}$, the electron moves as a free particle with momentum $\not{p}_1 + \not{Q}$ until it emits a photon of polarization $\not{e}$. We then ask: Is the electron in state u_2? For the Coulomb potential

$$\not{V}(Q) = (4\pi Z e^2/Q^2)\,\delta(Q_4)\,\gamma_t = v(Q)\,\delta(Q_4)\,\gamma_t$$

(see Momentum Representation, Lecture 18) in a coordinate system in which the nucleus does not move. [For potential other than Coulomb, use appropriate v(Q), the Fourier transform of the space dependence of the potential.] Rationalizing the denominator of the matrix,†

$$M = (-i)(4\pi e^2)^{1/2}\, v(Q) \left[\not{e} \frac{\not{p}_1 + \not{Q} + m}{-2\mathbf{p}_1 \cdot \mathbf{Q} - \mathbf{Q}^2} \gamma_t + \gamma_t \frac{\not{p}_2 - Q + m}{2\mathbf{p}_2 \cdot \mathbf{Q} - \mathbf{Q}^2} \not{e} \right] \tag{22-4}$$

The outgoing photon can be polarized in either of two directions, and the incoming and outgoing electron each have two possible spin states. The various matrix elements can be worked out using Table 13-1 exactly as was done in deriving the Klein-Nishina cross section in Lecture 20. Nothing new is involved, so we omit the details. After (1) summing over photon polarizations, (2) summing over outgoing electron spin states, and (3) averaging over incoming electron spin states, the following differential cross section is obtained:

$$†(\not{p}_1 + \not{Q} - m)(\not{p}_1 + \not{Q} + m) = p_1^2 + 2p_1 \cdot Q + Q^2 - m^2 = 2p_1 \cdot Q + Q^2$$
$$= -2\mathbf{p}_1 \cdot \mathbf{Q} + \mathbf{Q}^2$$
$$= 2\mathbf{p}_1 \cdot \mathbf{Q} - \mathbf{Q}^2 \qquad Q_4 = 0$$

$$d\sigma = \frac{1}{2\pi}\left(\frac{Ze^2}{Q^2}\right)^2 e^2 \frac{d\omega}{\omega}\frac{p_2}{p_1} \sin\theta_2\, d\theta_2 \sin\theta_1\, d\theta_1\, d\phi$$

$$\times \left\{\frac{p_2^2 \sin^2\theta_2 (4E_1^2 - Q^2)}{(E_2 - p_2\cos\theta_2)^2} + \frac{p_1^2\sin^2\theta_1 (4E_2^2 - Q^2)}{(E_1 - p_1\cos\theta_1)^2}\right.$$

$$\left. - \frac{2p_1p_2 \sin\theta_1 \sin\theta_2 \cos\phi\,(4E_1E_2 - Q^2 + 2\omega^2) - 2\omega^2(p_2^2\sin^2\theta_2 + p_1^2\sin^2\theta_1)}{(E_2 - p_2\cos\theta_2)(E_1 - p_1\cos\theta_1)}\right\}$$

(22-5)

An approximate expression with a simple interpretation in terms of the Coulomb elastic scattering cross section can be obtained when the photon energy is small (small compared to rest mass of electron but large compared to electron binding energies). Writing the matrix (22-3) in terms of $\not{q}$ instead of $\not{Q}$,

$$M = (-i)(4\pi e^2)^{1/2}\left[\not{e}\frac{1}{\not{p}_2 + \not{q} - m}\not{V}(Q) + \not{V}(Q)\frac{1}{\not{p}_1 - \not{q} - m}\not{e}\right]$$

$$= (-i)(4\pi e^2)^{1/2}\left[\not{e}\frac{\not{p}_2 - \not{q} + m}{+2p_2\cdot q}\not{V}(Q) + \not{V}(Q)\frac{\not{p}_1 - \not{q} + m}{-2p_1\cdot q}\not{e}\right]$$

using the relationships $\not{e}\not{p}_2 = -\not{p}_2\not{e} + 2e\cdot p_2$, $\not{p}_1\not{e} = -\not{e}\not{p}_1 + 2e\cdot p_1$, and neglecting $\not{q}$ in the numerator, since it is small, this becomes

$$M \approx (-i)(4\pi e^2)^{1/2}\, v(Q)\left[\frac{-\not{p}_2\not{e}\gamma_t + 2e\cdot p_2\gamma_t + m\not{e}\gamma_t}{2p_2\cdot q}\right.$$

$$\left. + \frac{-\gamma_t\not{e}\not{p}_1 + 2p_1\cdot e\gamma_t + m\not{e}\gamma_t}{-2p_1\cdot q}\right]\delta(Q_4)$$

$$= (-i)(4\pi e^2)^{1/2}\, v(Q)\left[\frac{e\cdot p_1}{q\cdot p_1} - \frac{e\cdot p_2}{q\cdot p_2}\right]\gamma_t\,\delta(Q_4)$$

where use is made of the fact that the matrix element of M between states u_2 and u_1 is to be calculated and $\tilde{u}_2\not{p}_2 = \tilde{u}_2 m$, $\not{p}_1 u_1 = m u_1$.

The cross section for photon emission can then be written

$$d\sigma = \frac{1}{v}\left[\frac{2\pi}{2E_1 2E_2}|v(Q)|^2\frac{E_2 p_2\, d\Omega_2}{(2\pi)^3}\right]\left[\frac{e^2\, d\omega\cdot d\Omega_\omega}{\pi\omega}\left(\frac{p_2\cdot e}{p_2\cdot\frac{q}{\omega}} - \frac{p_1\cdot e}{p_1\cdot\frac{q}{\omega}}\right)^2\right]$$

The first bracket is the probability of transition for elastic scattering (see Lecture 16), so the last bracket may be interpreted as the probability of photon emission in frequency interval $d\omega$ and solid angle $d\Omega_\omega$ if there is elastic scattering from momentum p_1 to p_2.

Problem: Calculate the amplitude for emission of two low-energy photons by the foregoing method. Neglect q's in the numerator but not in the denominator.

Answer: Another factor, similar to that in the preceding equations, is obtained for the extra photon.

PAIR PRODUCTION

It is easily shown that a single photon of energy greater than 2m cannot create an electron positron pair without the presence of some other means of conserving momentum and energy. Two photons could get together and create a pair, but the photon density is so low that this process is extremely unlikely. A photon can, however, create a pair with the aid of a field, such as that of a nucleus, to which it can impart some momentum. As with bremsstrahlung, there are two indistinguishable ways in which this can happen: (a) The incoming photon creates a pair and subsequently the electron interacts with the field of the nucleus; or (b) the photon creates a pair and the positron interacts with the field of the nucleus. The diagrams for these alternatives are shown in Fig. 22-3. The arrows in the diagram indicate that

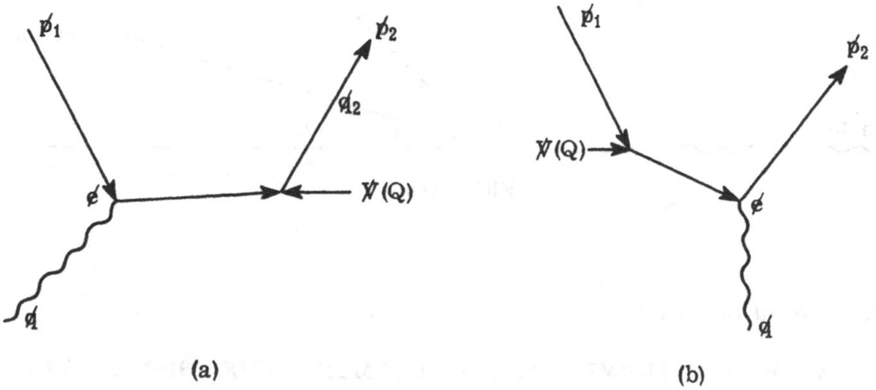

FIG. 22-3

$\not{p}_1$ is the positron momentum and $\not{p}_2$ is the electron momentum. Notice that, with respect to the directions that the arrows point (and without regard to direction of increasing time), these diagrams look exactly like those for the bremsstrahlung process: Starting with $\not{p}_1$ in case (a), the particle is first scattered by the Coulomb potential and then by the photon; in case (b) the order of the events is reversed. The difference between pair production and bremsstrahlung, when the direction of time is taken into account, is (1) $\not{p}_1$ is a positron state (an electron traveling backward in time), and (2) the photon $\not{q}$ is absorbed rather than emitted. As a result, the bremsstrahlung matrix elements can be used for this process if $\not{p}_1$ is replaced by $-\not{p}_+$ and $\not{q}$ by $-\not{q}$.

The $\not{p}_+$ is then the positron momentum and $\not{q}$ is the momentum of the absorbed photon. The density of final states is different, of course, since the particles in the final state are now a positron and electron. Thus

$$d\sigma = (1/2\pi)(Ze^2/Q^2)^2\, e^2\, (p_+ p_- \sin\theta_+\, d\theta_+\, \sin\theta_-\, d\theta_-\, d\phi/\omega^3) \times \{\ \} \tag{22-6}$$

where the braces are the same as for bremsstrahlung, Eq. (22-5), except for the following substitutions:

p_- for p_2	$-\theta_-$ for θ_2	E_- for $\overline{E}_+$
$-p_+$ for p_1	$-\theta_+$ for θ_-	$-E_+$ for E_1
		$-\omega$ for ω

Figure 22-4 defines the angles (ϕ = angle between electron-photon plane and positron-photon plane).

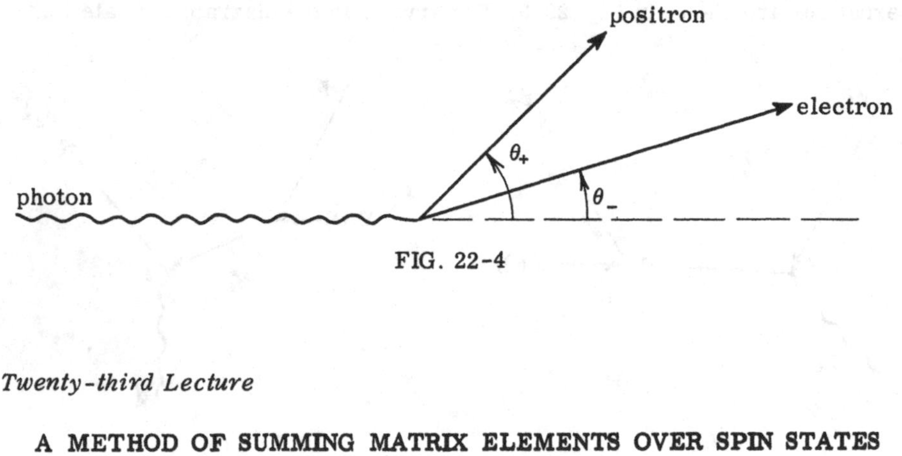

FIG. 22-4

Twenty-third Lecture

A METHOD OF SUMMING MATRIX ELEMENTS OVER SPIN STATES

By using current methods of computing cross sections, one first arrives at a cross section for "polarized" electrons, that is, electrons with definite incoming and outgoing spin states. In practice it is common that the incoming beam will be "unpolarized" and the spins of the outgoing particles will be unobserved. In this case, one needs the cross section obtained from that for "polarized" electrons by summing probabilities over final spin states and averaging this sum over initial spin states. This is the correct process since the final spin states do not interfere and there is equal probability of initial spin in either direction. Formally, if

$$\sigma \sim |(\tilde{u}_2 M u_1)|^2$$

one needs

$$\sigma \sim \frac{1}{2} \sum_{\text{spins 1}} \sum_{\text{spins 2}} |(\tilde{u}_2 M u_1)|^2 \tag{23-1}$$

where $\sum_{\text{spins 2}}$ means the sum over final spin states for only one sign of the the energy, that is, over only two of the four possible eigenstates. Similarly, $\sum_{\text{spins 1}}$ is the sum over initial spins for one sign of the energy. The purpose now is to develop a simple method for obtaining these sums.

In accordance with the usual rule for matrix multiplication, the following is true:

$$\sum_{\text{all } u_1} (\tilde{u}_2 A u_1)(\tilde{u}_1 B u_2) = 2m(\tilde{u}_2 AB u_2) \tag{23-2}$$

where A and B are any operators or matrices, the 2m factor on the right arises from the normalization $\tilde{u}u = 2m$, and the sum is over *all* eigenstates represented by u_1. But the states u, which we want in Eq. (23-1) are not all states, just those satisfying $\not{p}_1 u_1 = m u_1$. That is, they belong to the eigenvalue $+m$ of the operator $\not{p}_1$. Since $\not{p}_1^2 = m^2$, $\not{p}_1$ also has the eigenvalue $-m$, that is, there are two more solutions of $\not{p}_1 u = -mu$ which, together with the two we wish in Eq. (23-1) bring the total to four. Let us call the latter "negative eigenvalue" states.

Now, if in Eq. (23-2) the matrix elements of B were zero in negative eigenvalue states, this would be the same as $\sum_{\text{spins 1}}$, that is, just over positive eigenvalue states. So consider

$$\sum_{\text{all } u_1} (\tilde{u}_2 A u_1)(\tilde{u}_1(\not{p}_1 + m)B u_2) = (\tilde{u}_2 A(\not{p}_1 + m)B u_2)2m$$

But

$$\tilde{u}_1(\not{p}_1 + m) = 0 \qquad \text{for negative eigenvalue states}$$

$$= \tilde{u}_1(2m) \qquad \text{for positive eigenvalue states}$$

so the preceding sum also equals

$$\sum_{\text{spins 1}} (\tilde{u}_2 A u_1) 2m (\tilde{u}_1 B u_2)$$

Cancelling the 2m factors, this gives

$$\sum_{\text{spins 1}} (\tilde{u}_2 A u_1)(\tilde{u}_1 B u_2) = (\tilde{u}_2 A(\not{p}_1 + m) B u_2)$$

$(\not{p}_1 + m)$ is called a projection operator for obvious reasons. Similarly it follows that

$$\sum_{\text{spins 2}} (\tilde{u}_2 X u_2) = \sum_{\text{all } u_2} (1/2m)(\tilde{u}_2(\not{p}_2 + m)Xu_2)$$

where X is again any matrix. Remembering the normalization $\tilde{u}_2 u_2 = 2m$, it is seen that the last sum is just the trace or spur of the matrix $(\not{p}_2 + m)X$. Note that the order of X and $\not{p}_2 + m$ is immaterial.

Finally, when one wants

$$\sum_{\text{spins 1}} \sum_{\text{spins 2}} |(\tilde{u}_2 M u_1)|^2$$

collection and specialization of the previous results is seen to give

$$1/2 \sum_{\text{spins 1}} \sum_{\text{spins 2}} |\tilde{u}_2 M u_1|^2 = 1/2 \sum_{\text{spins 1}} \sum_{\text{spins 2}} (\tilde{u}_2 M u_1)(\tilde{u}_1 \tilde{M} u_2)$$

$$= 1/2 \, \mathrm{Sp}[(\not{p}_2 + m) M (\not{p}_1 + m) \tilde{M}] \tag{23-3}$$

where the last notation means the spur of the matrix in the brackets. It is true whether $\not{p}_1$, $\not{p}_2$ represent electrons or positrons.

The following list of the spurs of several frequently encountered matrices may be verified easily:

$$\mathrm{Sp}[1] = 4 \qquad \mathrm{Sp}[\gamma_\mu] = 0 \qquad \mathrm{Sp}[xy] = \mathrm{Sp}[yx]$$

$$\mathrm{Sp}[x+y] = \mathrm{Sp}[x] + \mathrm{Sp}[y]$$

$$\begin{aligned} \mathrm{Sp}[\gamma_\nu \gamma_\mu] &= 0 \qquad \text{if } \mu \neq \nu \\ &= +4 \qquad \text{if } \mu = \nu = 4 \\ &= -4 \qquad \text{if } \mu = \nu = 1, 2, 3 \end{aligned}$$

$$\mathrm{Sp}[\not{a}\not{b}] = 1/2 \, \mathrm{Sp}[\not{a}\not{b} + \not{b}\not{a}] = \mathrm{Sp}[a \cdot b] = 4\, a \cdot b$$

$$\mathrm{Sp}[\not{a}\not{b}\not{c}] = 0$$

It is also true that the spur of the product of any *odd* number of daggered operators is zero.

$$\mathrm{Sp}[(\not{p}_1 + m_1)(\not{p}_2 - m_2)] = \mathrm{Sp}[\not{p}_1\not{p}_2] + \mathrm{Sp}[m_1\not{p}_2 - \not{p}_1 m_2 - m_1 m_2]$$

$$= 4(p_1 \cdot p_2 - m_1 m_2) \tag{23-4}$$

$$\begin{aligned} &\mathrm{Sp}[(\not{p}_1 + m_1)(\not{p}_2 - m_2)(\not{p}_3 + m_3)(\not{p}_4 - m_4)] \\ &\quad = 4(p_1 \cdot p_2 - m_1 m_2)(p_3 \cdot p_4 - m_3 m_4) - 4(p_1 \cdot p_3 - m_1 m_3) \\ &\quad\quad \times (p_2 \cdot p_4 - m_2 m_4) + 4(p_1 \cdot p_4 - m_1 m_4)(p_2 \cdot p_3 - m_2 m_3) \end{aligned} \tag{23-5}$$

As an example, the case of Coulomb scattering will be "treated" using this technique. The cross section for polarized electrons was previously found to be

$$\sigma = (Z^2e^4/Q^4)\,|(\tilde{u}_2\gamma_t u_1)|^2$$

Therefore, since $\tilde{\gamma}_t = \gamma_t$, the cross section for unpolarized electrons is, by Eq. (23-3),

$$\sigma_{\text{unpol}} = 1/2\ (Z^2e^4/Q^4)\ \text{Sp}[(\not{p}_2 + m)\gamma_t(\not{p}_1 + m)\gamma_t]$$

The spur can be evaluated immediately from Eq. (23-5) with $m_2 = m_4 = 0$ and $\not{p}_2 = \not{p}_4 = \gamma_t$. Another way is: Since $\gamma_t\not{p}_1 = 2E_1 - \not{p}_1\gamma_t$, it is seen that

$$(\not{p}_2 + m)\gamma_t(\not{p}_1 + m)\gamma_t = (\not{p}_2 + m)(2E_1\gamma_t - \not{p}_1 + m)$$

Using a few of the formulas listed previously, the spur of this matrix is seen to be

$$-4p_1 \cdot p_2 + 8E_1E_2 + 4m^2$$

But $p_1 \cdot p_2 = E_1E_2 - \mathbf{p}_1 \cdot \mathbf{p}_2$, $\mathbf{p}_1 \cdot \mathbf{p}_2 = p^2 \cos\theta$, and $E_1 = E_2$, so this is

$$4E^2 + 4m^2 + 4p^2 \cos\theta$$

Also $m^2 = E^2 - p^2$, so that finally the cross section becomes

$$\sigma_{\text{unpol}} = 1/2\ (Z^2e^4/Q^4)[8E^2 + 4p^2(\cos\theta - 1)]$$

$$= (4Z^2e^4/Q^4)\,E^2[1 - v^2 \sin^2(\theta/2)]$$

where $v^2 = p^2/E^2$. This is the same cross section obtained previously by other methods.

EFFECTS OF SCREENING OF THE COULOMB FIELD IN ATOMS

The cross sections for the pair production and bremsstrahlung processes contained the factor $[V(\mathbf{Q})]^2$, where $V(\mathbf{Q})$ is the momentum representation of the potential; that is,

$$V(\mathbf{Q}) = \int V(R) \exp(-i\mathbf{Q} \cdot \mathbf{R})\, d^3\mathbf{R}$$

which for a Coulomb potential is

$$V(\mathbf{Q}) = 4\pi Ze^2/Q^2$$

where $\mathbf{Q}$ is the momentum transferred to the nucleus or $\mathbf{p}_1 - \mathbf{p}_2 - \mathbf{q}$.

Clearly V(**Q**) gets large as **Q** gets small. The minimum value of **Q** occurs when all three momenta are lined up (Fig. 23-1):

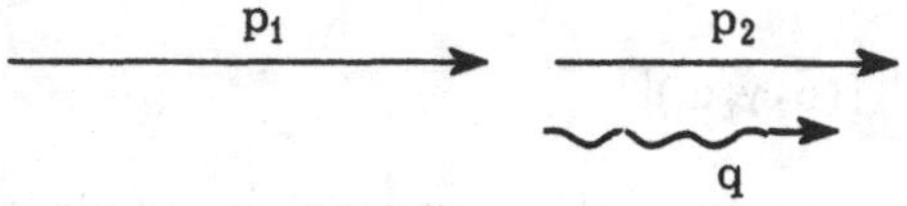

FIG. 23-1

$$|Q_{min}| = \mathbf{p}_1 - \mathbf{p}_2 - \mathbf{q}$$

$$= |\mathbf{p}_1| - |\mathbf{p}_2| - (E_1 - E_2)$$

For very high energies $E \gg m$,

$$E - p \approx m^2/2E$$

so that in this case

$$Q_{min} = (m^2/2)[(1/E_2) - (1/E_1)] \approx m^2q/2E_1 E_2$$

From this it is seen that $Q_{min} \to 0$ as $E_1 \to \infty$. This shows clearly why the cross sections for pair production and bremsstrahlung go up with energy.

From the integral expression for V(Q) it is seen that the main contribution to the integral comes when $R \sim 1/Q$. So as Q becomes small the important range of R gets large. It is in this way that screening of the Coulomb field becomes effective. The value of $1/Q_{min}$ for a contemplated process can be estimated from the foregoing formula. The atomic radius is given roughly by $a_0 Z^{-1/3}$, where a_0 is the Bohr radius. Thus if

$$R_{eff} = 1/Q_{min} > a_0 Z^{-1/3}$$

or, what is the same,

$$E_1E_2/q > 1/2\ (137)\ mZ^{-1/3}$$

then screening effect will be important, and vice versa for the opposite inequalities. If from this estimate screening would appear to be important, one should use the screened Coulomb potential. It gives the result

$$V(Q) = (4\pi e^2/Q^2)[Z - F(Q)]$$

where F(Q) is the atomic structure factor given by

$$F(Q) = \int n(R) \exp(-i\mathbf{Q} \cdot \mathbf{R})\, d^3R$$

and n(R) is the electron density as a function of R.

Twenty-fourth Lecture

Problem: In discussing bremsstrahlung it was found that the cross section for emission of a low-energy photon can be approximated as

$$\sigma = \sigma_0 e^2 4\pi \, d\Omega \, (d\omega/\pi\omega)[p_2 \cdot e/p_2 \cdot (q/\omega) - p_1 \cdot e/p_1 (q/\omega)]^2 \tag{24-1}$$

where σ_0 is the scattering cross section (neglecting emission). Now consider an energetic Compton scattering in which a third, weak photon is emitted. The three diagrams are shown in Fig. 24-1.

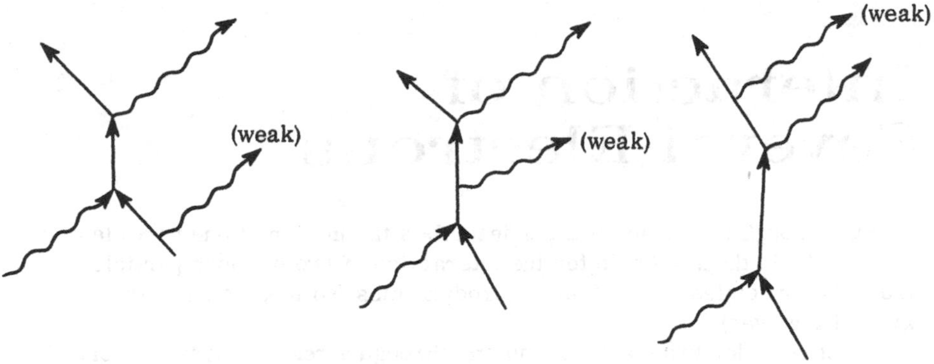

FIG. 24-1

Show that the cross section for this effect is given by Eq. (24-1), with the Klein-Nishina formula replacing σ_0. (Remember to assume q small.)

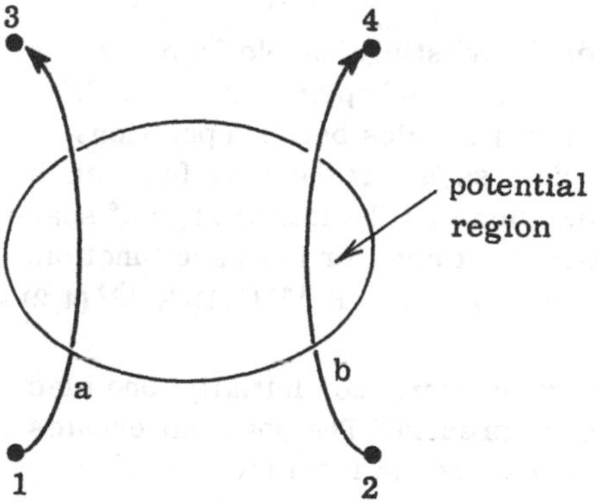

FIG. 24-2

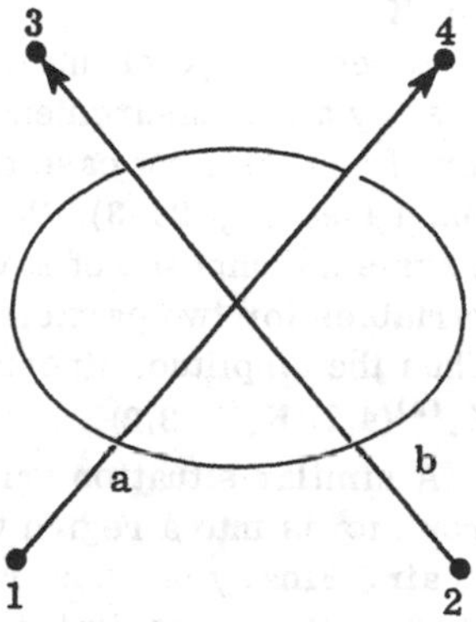

FIG. 24-3

Interaction of Several Electrons

Even though the Dirac equation describes the motion of one particle only, we can obtain the amplitude for the interaction of two or more particles from the principles of quantum electrodynamics (so long as nuclear forces are not involved).

First consider two electrons moving through a region where a potential is present and assume that they do not interact with one another (see Fig. 24-2). The amplitude for electron a moving from $1 \to 3$, while electron b moves from $2 \to 4$ is given the symbol K(3,4;1,2). If it is assumed that no interaction between electrons takes place, then K can be written as the product of kernels $K_+^{(a)}(3,1)\, K_+^{(b)}(4,2)$, where the superscript means that $K_+^{(a)}$ operates only on those variables describing particle a, and similarly for $K_+^{(b)}$.

A second type of interaction gives a result indistinguishable from the first by any measurement in accordance with the Pauli principle. This differs from the first case by the interchange of particles between positions 3 and 4 (see Fig. 24-3). Now the Pauli principle says that the wave function of a system composed of several electrons is such that the interchange of space variables for two particles results in a change of sign for the wave function. Thus the amplitude (including both possibilities) is $K = K_+^{(a)}(3,1)\, K_+^{(b)}(4,2) - K_+^{(a)}(4,1)\, K_+^{(b)}(3,2)$.

A similar situation arises in the following occurrence. Initially, one electron moves into a region where a potential is present. The potential creates a pair. Finally one positron and two electrons emerge from the region. There are two possibilities for this occurrence, as shown in Fig. 24-4. Again, the total amplitude for the occurrence is the difference between the amplitudes for the two possibilities.

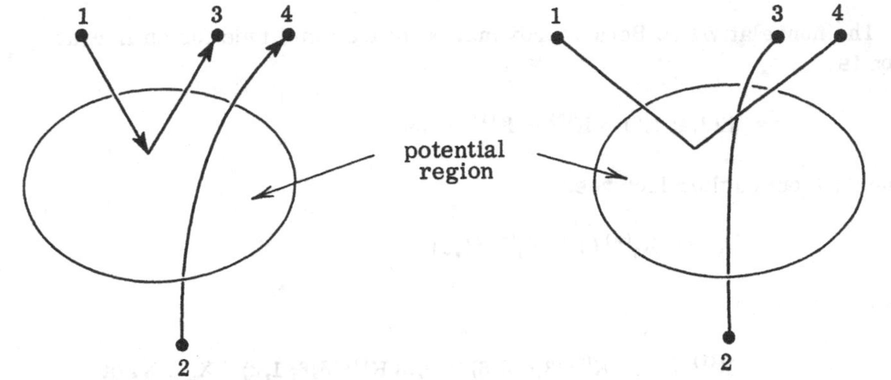

FIG. 24-4

The probability of this occurrence, or the previous, or any other similar occurrence is given by the absolute square of the amplitude times the number P_V. The P_V is actually the probability that a vacuum remains a vacuum; because of the possibility of pair production, it is not unity. The P_V can be computed by making a table of the probabilities of starting with nothing and ending with various numbers of pairs, as is shown in Table 24-1.

TABLE 24-1

Final number of pairs	Probability
0	$P_V\, 1^2$
1	$P_V\, \lvert K_+(2,1)\rvert^2$
2	$P_V\, \lvert K_+(3,1)\, K_+(4,2) - K_+(4,1)\, K_+(3,2)\rvert^2$
3	etc.
etc.	

The sum of all these probabilities must equal unity, and P_V is determined from this equation. The magnitude of P_V depends on the potential present. So the "probabilities" taken as merely the squares of amplitudes (that is, omitting the P_V factor) are actually relative probabilities for various occurrences in a given potential.

Use of $\delta_+(s^2)$. For the present, the existence of more than one possibility for an occurrence (the Pauli principle) will be neglected. The total amplitude can always be derived from one by interchanging the proper space variables, making the corresponding changes in sign, and summing all the amplitudes so obtained.

The nonrelativistic Born approximation to the amplitude for an interaction is

$$K(3,4;\ 1,2) = K^{(0)} + K^{(1)}$$

where, from earlier lectures,

$$K^{(0)} = K_0^{(a)}(3,1)\, K_0^{(b)}(4,2)$$

and

$$K^{(1)} = -i \int K^{(0)}(3,4;5,6)\, V(5,6)\, K^{(0)}(5,6;\ 1,2)\, d^3X_5\, d^3X_6\, dt_5$$

Note that $t_5 = t_6$ since a nonrelativistic interaction affects both particles simultaneously. The potential for the interaction is the Coulomb potential

$$V(5,6) = e^2/r_{5,6}$$

Separate variables may be used for t_5 and t_6, if the function $\delta(t_5 - t_6)$ is included as a factor. Then

$$K^{(1)} = -i \iint K_0(3,5)\, K_0(4,6)(e^2/r_{5,6})\delta(t_5 - t_6)\, K_0(5,1)\, K_0(6,2) \times d\tau_5\, d\tau_6$$

where the differential $d\tau$ includes both space and time variables. It is conceivable that the relativistic kernel could be obtained by substituting K_+ for K_0, and introducing the idea of a retarded potential by replacing $\delta(t_5 - t_6)$ by $\delta(t_5 - t_6 - r_{5,6})$. However this δ function is not quite right. Its Fourier transform contains both positive and negative frequencies, whereas a photon has only positive energy. Thus

$$\delta(X) = \int_{-\infty}^{\infty} \exp(-i\omega X)\, d\omega/2\pi$$

To correct this, define the function

$$\delta_+(X) = \int_0^{\infty} \exp(-i\omega X)\, d\omega/\pi$$

which contains only positive energy. The value of the function is determined by the integral. Thus,

$$\begin{aligned} \delta_+(X) &= \lim_{\epsilon \to 0} (1/\pi i)(X - i\epsilon) \\ &= \delta(X) + (1/\pi i)(\text{principal value } 1/X) \end{aligned}$$

Abbreviating $t_5 - t_6 \equiv t$ and $r_{5,6} = r$, and taking account of the fact that both $t_5 \leq t_6$ and $t_5 \geq t_6$ are possible, the retarded potential is

$$V(5,6) = (e^2/2r)[\delta_+(t-r) + \delta_+(-t-r)]$$

Exercises: (1) Show that

$$(1/2r)[\delta_+(t-r) + \delta_+(-t-r)] = \delta_+(t^2 - r^2)$$

Defining $t^2 - r^2$ as $s_{5,6}{}^2$, a relativistic invariant, the potential is $e^2\delta_+(s_{5,6}{}^2)$. Another term which must be included is the magnetic interaction, proportional to $-V_a \cdot V_b$. In the notation used for the Dirac equation, this product is $-\alpha_a \cdot \alpha_b$. It will be found convenient to express this in the equivalent form $-(\beta\alpha)_a \cdot (\beta\alpha)_b$, and in this notation the retarded Coulomb potential is proportional to $\beta_a\beta_b$. These β's come from the use of the relativistic kernel. Thus the complete potential for the interaction becomes

$$e^2\delta_+(s_{5,6}{}^2)[\beta_a \cdot \beta_b - (\beta\alpha)_a \cdot (\beta\alpha)_b] = e^2\delta(s_{5,6}{}^2)\,\gamma_\mu{}^{(a)}\gamma_\mu{}^{(b)}$$

and then the first-order kernel is

$$\begin{aligned} K^{(1)}(3,4;1,2) &= -ie^2 \int\int K_+{}^{(a)}(3,5)\, K_+{}^{(b)}(4,6)\,\gamma_\mu{}^{(a)}\gamma_\mu{}^{(b)} \\ &\qquad \times \delta_+(s_{5,6}{}^2)\, K_+{}^{(a)}(5,1)\, K_+{}^{(b)}(6,2)\, d\tau_5\, d\tau_6 \\ &= -ie^2 \int\int [K_+(3,5)\,\gamma_\mu K_+(5,1)]_a\, \delta_+(s_{5,6}{}^2) \\ &\qquad \times [K_+(4,6)\,\gamma_\mu K_+(6,2)]_b\, d\tau_5\, d\tau_6 \qquad (24\text{-}2) \end{aligned}$$

Here the superscript on γ_μ indicates on which set of variables the matrix operates, just as for the superscripts on K_+.

The occurrence represented by this kernel can be diagrammed as in Fig. 24-5. This represents the exchange of a virtual photon be-

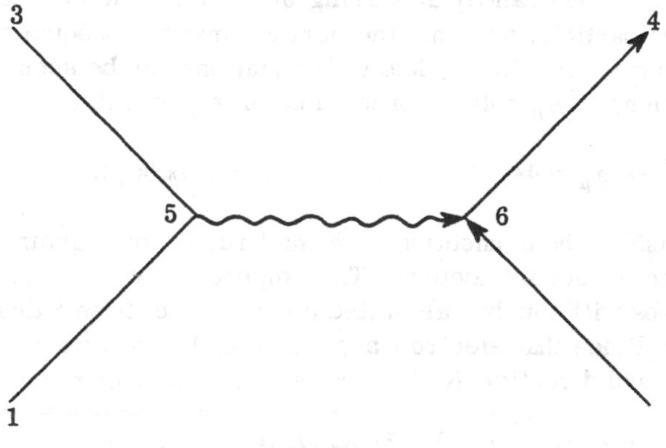

FIG. 24-5

tween the electrons. The virtual photon can be polarized in any one of the four directions, t, x, y, z. Summation over these four possibilities is indicated by the repeated index of $\gamma_\mu \gamma_\mu$. The integral expression for the kernel, Eq. (24-2), implies that the amplitude for a photon to go from $5 \to 6$ (or from $6 \to 5$ depending on timing) is $\delta_+(s_{5,6}^2)$. Equation (24-2) can be taken as another statement of the fundamental laws of quantum electrodynamics.

(2) Show that

$$\delta_+(s^2) = -4\pi \int [\exp(-ik \cdot X)]\, d^4k/(k^2 + i\epsilon)(2\pi)^4$$

Thus, in momentum space,

$$\delta_+(s^2) \to -4\pi/k^2$$

Twenty-fifth Lecture

DERIVATION OF THE "RULES" OF QUANTUM ELECTRODYNAMICS

From the results of the last lecture, it is evident that the laws of electrodynamics could be stated as follows: (1) The amplitude to emit (or absorb) a photon is $e\gamma_\mu$, and (2) the amplitude for a photon to go from 1 to 2 is $\delta_+(s_{1,2}^2)$, where

$$\delta_+(s_{1,2}^2) = -4\pi \int e^{\frac{-ik \cdot (x_2 - x_1)}{k^2 + i\epsilon}} \frac{d^4k}{(2\pi)^4} \qquad (25\text{-}1)$$

$$= -4\pi/(k^2 + i\epsilon)$$

in momentum representation. It is interesting to note that $\delta_+(s_{1,2}^2)$ is the same as $I_+(s_{1,2}^2)$, the quantity appearing in the derivation of the propagation kernel of a free particle, with m, the particle mass, set equal to zero. A more direct connection with the Maxwell equations can be seen by writing the wave equation, $\Box^2 A_\mu = 4\pi J_\mu$ in momentum representation,

$$-k^2 a_\mu = 4\pi j_\mu \qquad \text{or} \qquad a_\mu = -(4\pi/k^2)\, j_\mu \qquad (25\text{-}2)$$

We now consider the connection with the "rules" of quantum electrodynamics given in the second lecture. The amplitude for a to emit a photon which b absorbs will now be calculated according to those rules (see Fig. 25-1). The amplitude that electron a goes from 1 to 5, emits a photon of polarization $\not{e}$ and direction **K**, then goes from 5 to 3 is given by

$$[K_+(3,5)\, \not{e} \sqrt{(4\pi e^2/2K)}\, \exp(-i\mathbf{K} \cdot \mathbf{r}_5) \exp(iKt_5)\, K_+(5,1)]_a$$

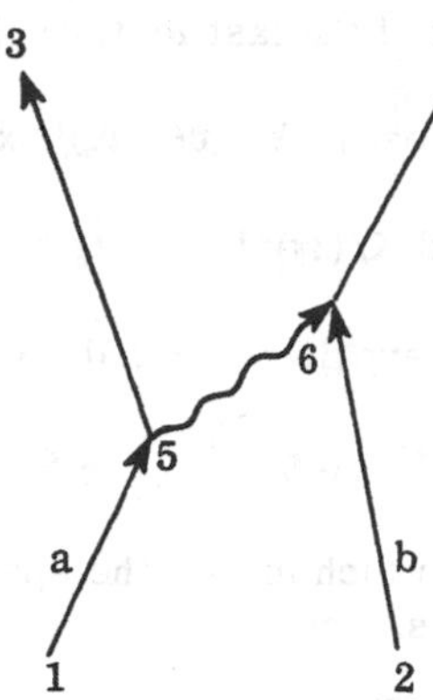

FIG. 25-1

whereas the amplitude that b goes from 2 to 6, absorbs a photon of polarization $\not{e}$ and direction **K** at 6, then goes from 6 to 4 is given by

$$[K_+(4,6)\ \not{e}\ \sqrt{(4\pi e^2/2K)}\ \exp(i\mathbf{K}\cdot\mathbf{r}_6)\exp(-iKt_6)\ K_+(6,2)]_b$$

The amplitude that both these processes occur, which is equivalent to b absorbing a's photon if $t_6 > t_5$ is just the product of the individual amplitudes. If a absorbs b's photon, the signs of all the exponentials in the preceding amplitudes are changed and t_6 must be less than t_5.

To obtain the amplitude that any photon is exchanged between a and b, it is necessary to integrate over photon direction, sum over possible photon polarizations, and integrate over t_5 and t_6, subject to the aforementioned restrictions. In summing over polarizations, $\not{e}$ will be replaced by γ_μ and a summation over μ will be taken. This amounts to summing over four directions of polarization, something that will be explained later. Thus

$$\begin{aligned}
\begin{Bmatrix}\text{Amp. for}\\ \text{photon}\\ a \rightarrow b\end{Bmatrix} &= 4\pi e^2 \sum_\mu \int \exp[-i\mathbf{K}\cdot(\mathbf{r}_5-\mathbf{r}_6)]\ \exp[iK(t_5-t_6)]\\
&\quad\times [K_+(3,5)\gamma_\mu K_+(5,1)]_a\ [K_+(4,6)\gamma_\mu K_+(6,2)]_b\\
&\quad\times (1/2K)[d^3K/(2\pi)^3]\ dt_5\ dt_6 \qquad t_6 > t_5\\
&= 4\pi e^2 \sum_\mu \int \exp[i\mathbf{K}\cdot(\mathbf{r}_5-\mathbf{r}_6)]\ \exp[-iK(t_5-t_6)]\\
&\quad\times [K_+(3,5)\gamma_\mu K_+(5,1)]_a\ [K_+(4,6)\gamma_\mu K_+(6,2)]_b\\
&\quad\times (1/2K)[d^3K/(2\pi)^3]\ dt_5\ dt_6 \qquad t_6 < t_5
\end{aligned} \tag{25-3}$$

Comparing this with the result of the last lecture, it must be that

$$\delta_+(s_{5,6}^2) = 4\pi \int \exp[-i\mathbf{K}\cdot(\mathbf{r}_5 - \mathbf{r}_6)] \exp[iK(t_5 - t_6)(1/2K)$$
$$\times [d^3K/(2\pi)^3] \qquad t_6 > t_5$$
$$= 4\pi \int \exp[i\mathbf{K}\cdot(\mathbf{r}_5 - \mathbf{r}_6)] \exp[-iK(t_5 - t_6)] (1/2K)$$
$$\times [d^3K/(2\pi)^3] \qquad t_6 < t_5$$

This can be written in a form which makes the space-time symmetry evident by using the Fourier transform

$$\exp(-iK|t|) = \int_{-\infty}^{\infty} [2iK/(\omega^2 - K^2 + i\epsilon)] \exp(-i\omega t)\, d\omega/2\pi$$

so that the foregoing equation becomes

$$\delta_+(s_{5,6}^2) = -4\pi \int \frac{\exp[-ik\cdot(x_5 - x_6)]}{k_4^2 - \mathbf{K}\cdot\mathbf{K} + i\epsilon} \frac{d^4k}{(2\pi)^4} \qquad (25\text{-}4)$$

and comparing this with the result of the last problem of Lecture 24 establishes that the rules given in Lecture 2 are consistent with relativistic electrodynamics developed in the last lecture.

ELECTRON-ELECTRON SCATTERING

The theory will now be used to obtain the electron-electron scattering cross section. The diagrams for the two indistinguishable processes are shown in Fig. 25-2.

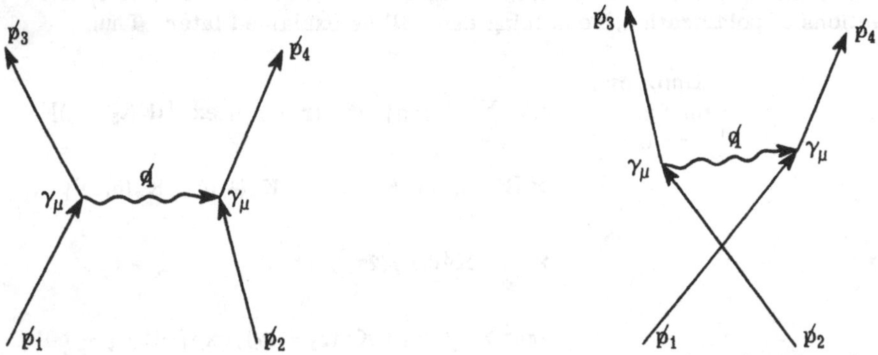

$q\!\!/ = p\!\!/_1 - p\!\!/_3 = p\!\!/_4 - p\!\!/_2$ R

$q\!\!/ = p\!\!/_2 - p\!\!/_3 = p\!\!/_4 - p\!\!/_1$ S

FIG. 25-2

The amplitude expressed in momentum representation is obtained as follows: Write Eq. (25-3) [with the aid of Eq. (25-4)] as

$$e^2 \sum_\mu \int [K_+(3,5)\gamma_\mu K_+(5,1)]_a \frac{-4\pi}{q^2} [K_+(4,6)\gamma_\mu K_+(6,2)]_b \frac{d^4q}{(2\pi)^4}$$
$$\times\, d\tau_5\, d\tau_6$$

Since electron state 1 is a plane wave of momentum $p\!\!/_1$ and electron state 3 is a plane wave of momentum $p\!\!/_3$, it is clear that in momentum representation the spinor part of the first bracket will become $(\tilde{u}_3\gamma_\mu u_1)$ and the spinor part of the second bracket will become $(\tilde{u}_4\gamma_\mu u_2)$. Integration over τ_5 and τ_6 produces the conservation laws given at the bottom of the diagrams. Dropping the integration over q puts the photon propagation in momentum representation directly. Thus the matrix element can be written

$$M = +i4\pi e^2 \left[\frac{(\tilde{u}_4\gamma_\mu u_2)(\tilde{u}_3\gamma_\mu u_1)}{(p\!\!/_1 - p\!\!/_3)^2} - \frac{(\tilde{u}_4\gamma_\mu u_1)(\tilde{u}_3\gamma_\mu u_2)}{(p\!\!/_4 - p\!\!/_2)^2}\right]$$

The first term comes from diagram R, the second from diagram S, and the summation over μ is implied. In the center-of-mass system, the probability of transition per second is

$$\text{Trans. prob./sec} = \sigma v_1 = \frac{2\pi}{(2E)^4}\,|M|^2\,\frac{E^2p^3\,d\Omega}{(2\pi)^3\,2Ep^2}$$

(see Density of Final States, Lecture 19). The method of Lecture 23 can be used to average over initial spin states and sum over final spin states. For example, the sums over spin states that result from $\tilde{R}$ by R matrices and $\tilde{R}$ by S plus R by $\tilde{S}$ matrices are

$$\tilde{R}R \rightarrow \frac{\mathrm{Sp}\,[(p\!\!/_4 + m)\gamma_\mu(p\!\!/_2 + m)\gamma_\nu]\,\mathrm{Sp}\,[(p\!\!/_3 + m)\gamma_\mu(p\!\!/_1 + m)\gamma_\nu]}{[(p\!\!/_1 - p\!\!/_3)^2]^2}$$

$$\tilde{R}S + R\tilde{S} \rightarrow -\frac{\mathrm{Sp}\,[(p\!\!/_4 + m)\gamma_\nu(p\!\!/_1 + m)\gamma_\mu(p\!\!/_3 + m)\gamma_\nu(p\!\!/_2 + m)\gamma_\mu]}{(p\!\!/_1 - p\!\!/_3)^2\,(p\!\!/_4 - p\!\!/_1)^2}$$

By judicious use of the spur relations given in Lecture 23 the following differential cross section is obtained (alternatively, Table 13-1 could be used to calculate M directly):

$$d\sigma = \frac{2e^4p\,d\Omega}{E^3}\left[\frac{4x^2 + 8x\cos\theta + 2(1-\cos^2\theta) + 4\cos\theta}{(1-\cos\theta)}\right.$$
$$\left. + \frac{4x^2 - 8x\cos\theta + 2(1-\cos^2\theta) - 4\cos\theta}{(1+\cos\theta)^2} - \frac{4(1+x)(x-3)}{(1-\cos\theta)(1+\cos\theta)}\right]$$

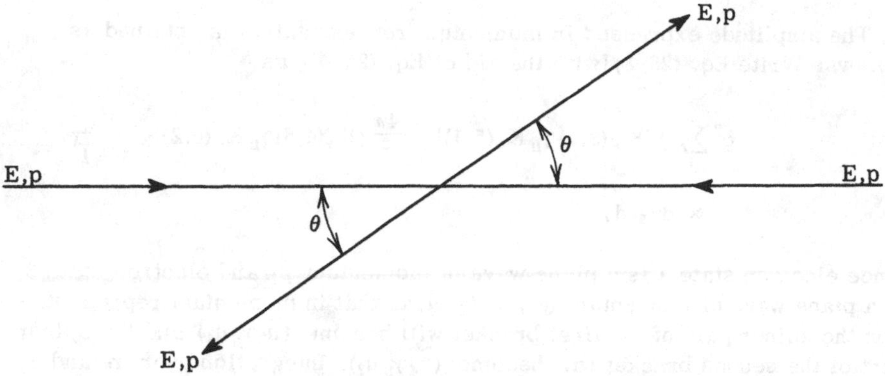

FIG. 25-3

where $x = E^2/p^2$. This is called Möller scattering (see Fig. 25-3).

Problems: (1) Calculate positron-electron scattering by the preceding method.

(2) Find the cross section for a μ meson to produce a knock-on electron. Assume that the μ meson satisfies the Dirac equation with $S = 1/2$ and no anomalous moment. Remember that the particles are distinguishable and hence there is no interchange of particles.

(3) Calculate the expected electron-proton scattering cross section assuming the proton has no structure but does have an anomalous moment. The Dirac equation for a proton is (see page 54)

$$(i\nabla\!\!\!/ + M - eA\!\!\!/ - (\mu/4M)\gamma_\mu\gamma_\nu\, F_{\mu\nu})\Psi = 0\dagger$$

Thus the perturbing potential can be taken as (see page 54)

$$eA\!\!\!/ + (e\mu/4M)\gamma_\mu\gamma_\nu(\nabla_\mu A_\nu - \nabla_\nu A_\mu)$$

and the coupling with a photon is

$$e\not{e} + (e\mu/4M)(\not{q}\not{e} - \not{e}\not{q}) \qquad \text{or} \qquad e\gamma_\mu + (e\mu/4M)(\not{q}\gamma_\mu - \gamma_\mu\not{q})$$

The Sum over Four Polarizations. In classical electrodynamics, longitudinal waves can always be eliminated in favor of transverse waves and an instantaneous Coulomb interaction. This is the approach used by Fermi (see Lecture 1), and it will now be demonstrated that the sum over four polarizations is also equivalent to transverse waves but plus an instantaneous Coulomb interaction. If instead of choosing space directions x, y, z, one direction parallel to **Q** (photon momentum) and two directions transverse to **Q** are taken, the matrix element can be written

† For the proton $\mu = 1.7896$.

$$M/-i4\pi e^2 = (\tilde{u}_4\gamma_t u_2)(1/q^2)(\tilde{u}_3\gamma_t u_1) - (\tilde{u}_4\gamma_Q u_2)(1/q^2)(\tilde{u}_3\gamma_Q u_1)$$
$$- \sum_{\text{2 tr. direc.}} (\tilde{u}_4\gamma_{tr} u_2)(1/q^2)(\tilde{u}_3\gamma_{tr} u_1)$$

where γ_Q is the γ matrix for the Q directions and γ_{tr} represents the γ matrix in either of the transverse directions. The matrix element of $\not{q} = q_4\gamma_t - Q\gamma_Q$ is zero in general (from the argument for gauge invariance).† Thus γ_Q can be replaced by $(q_4/Q)\gamma_t$ with the result

$$\frac{M}{4\pi e^2} = (\tilde{u}_4\gamma_t u_2)\frac{1}{q^2}\left(1-\frac{q_4{}^2}{Q^2}\right)(\tilde{u}_3\gamma_t u_1) - \sum_{1,2}(\tilde{u}_4\gamma_{tr} u_2)\frac{1}{q^2}(\tilde{u}_3\gamma_{tr} u_1)$$
$$= -(\tilde{u}_4\gamma_t u_2)\frac{1}{Q^2}(\tilde{u}_3\gamma_t u_1) - \sum_{1,2}(\tilde{u}_4\gamma_{tr} u_2)\frac{1}{q^2}(\tilde{u}_3\gamma_{tr} u_1)$$

Now $1/Q^2$ represents a Coulomb field in momentum space and γ_t is the fourth component of the current density or charge, so that the first term represents a Coulomb interaction while the second term contains the interaction through transverse waves.

---

† In our special case, it is easy to see directly, for example,

$$(\tilde{u}_4\not{q}u_2) = (\tilde{u}_4(\not{p}_2 - \not{p}_4)u_2) = (\tilde{u}_4\not{p}_2u_2) - (\tilde{u}_4\not{p}_4u_2)$$
$$= m(\tilde{u}_4u_2) - m(\tilde{u}_4u_2) = 0$$

Discussion and Interpretation of Various "Correction" Terms

Twenty-sixth Lecture

In many processes the behavior of electrons in the quantum-electrodynamic theory turns out to be the same as predicted by simpler theories save for small "correction" terms. It is the purpose of the present lecture to point out and discuss a few such cases.

ELECTRON-ELECTRON INTERACTION

The simplest diagrams for the interaction are shown in Fig. 26-1. The amplitude for the process has been found to be proportional, in momentum

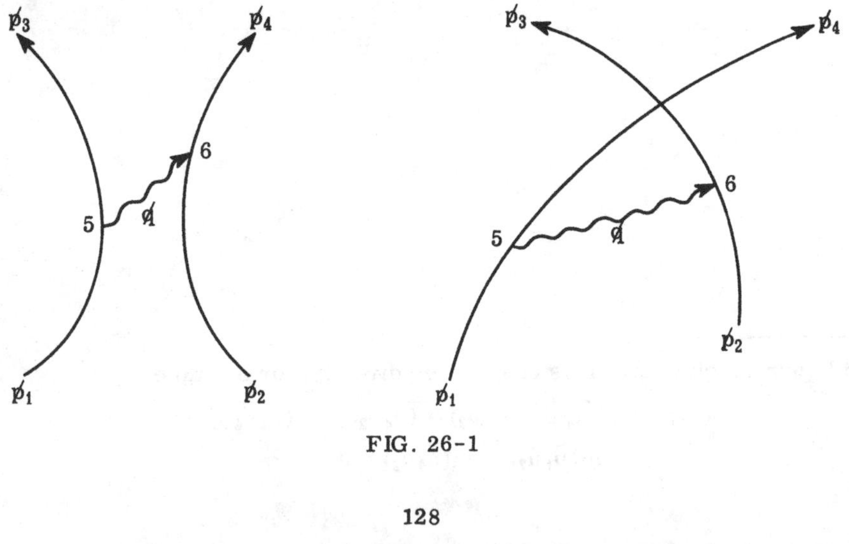

FIG. 26-1

representation, to

$$(\tilde{u}_3\gamma_\mu u_1)(\tilde{u}_4\gamma_\mu u_2)/q^2$$

where $q \equiv (\mathbf{Q}, q_4)$ and $\mathbf{Q}$ is the momentum exchanged by the two electrons. Also, since $\not{q} = \not{p}_1 - \not{p}_3$ it follows that

$$(\tilde{u}_3 \not{q} u_1) = (\tilde{u}_3 (\not{p}_1 - \not{p}_3) u_1) = 0$$

From this identity it was deduced in the last lecture that the amplitude for the process as just given is equivalent to

$$[-(\tilde{u}_3\gamma_t u_1)(\tilde{u}_4\gamma_t u_2)/Q^2] - \sum_{1,2} (\tilde{u}_3\gamma_{tr} u_1)(\tilde{u}_4\gamma_{tr} u_2)/q^2$$

By taking the Fourier transform of the first term, it can be seen that it is the momentum representation of a pure, instantaneous Coulomb potential. The second term then constitutes a correction to the simple Coulomb interaction. In it γ_{tr} denotes the γ's for two directions transverse to the direction of $\mathbf{Q}$.

For slow electrons, the correction to the Coulomb potential may be simplified and interpreted in a simple manner. Note that in this case

$$\mathbf{Q} = \mathbf{p}_1 - \mathbf{p}_3$$

and

$$q_4 = E_1 - E_2 \approx [m + (p_1^2/2m)] - [m + (p_3^2/2m)] = (p_1^2 - p_3^2)/2m$$

$$= [(\mathbf{p}_1 + \mathbf{p}_3)/2m](\mathbf{p}_1 - \mathbf{p}_3) \approx \mathbf{v}(\mathbf{p}_1 - \mathbf{p}_3)$$

so that $q_4^2 \sim v^2Q^2$ and q^2 in the denominator can be replaced by $-Q^2$ with small error. (In the C.G. system, $q_4 = 0$ exactly.) The correction term becomes

$$+\sum_{1,2} (\tilde{u}_3\gamma_{tr} u_1)(\tilde{u}_4\gamma_{tr} u_2)/Q^2$$

but

$$(\tilde{u}_3\gamma_{tr} u_1) \equiv u_3^* \boldsymbol{\alpha}_{tr} u_1$$

It is recalled that $u \equiv \begin{pmatrix} u_a \\ u_b \end{pmatrix}$, where u_a is the large part and u_b the small part and that in the nonrelativistic approximation

$$u_b \approx (1/2m)(\boldsymbol{\sigma} \cdot \boldsymbol{\Pi})u_a$$

Also, since

$$\boldsymbol{\alpha} = \begin{pmatrix} 0 & \boldsymbol{\sigma} \\ \boldsymbol{\sigma} & 0 \end{pmatrix}$$

it follows that (taken between positive energy states)

$$u_3^* \,\boldsymbol{\alpha}_{tr}\, u_1 = \widehat{u_{3a}^* u_{3b}^*} \begin{pmatrix} 0 & \boldsymbol{\sigma} \\ \boldsymbol{\sigma} & 0 \end{pmatrix}_{tr} \begin{pmatrix} u_{1a} \\ u_{1b} \end{pmatrix} = (u_{3a}^* \boldsymbol{\sigma}\, u_{1b} + u_{3b}^* \boldsymbol{\sigma}\, u_{1a})_{tr}$$

$$= 1/2m\,[u_{3a}^* \boldsymbol{\sigma}\,(\boldsymbol{\sigma}\cdot\Pi_1) + (\boldsymbol{\sigma}\cdot\Pi_3)\boldsymbol{\sigma}\, u_{1a}]_{tr}$$

In free space $\Pi \equiv p$, so the x component, for example, of the foregoing matrix is

$$\sigma_x(\sigma_x p_{1x} + \sigma_y p_{1y} + \sigma_z p_{1z}) + (\sigma_x p_{3x} + \sigma_y p_{3y} + \sigma_z p_{3z})\sigma_x$$

$$= (p_1 + p_3)_x + i\,[\sigma_z (p_1 - p_3)_y - \sigma_y (p_1 - p_3)_z]$$

where the commutation relations for the σ's have been used. From this it is easily seen that the amplitude for the correction to the Coulomb potential may be written altogether in the form

$$\sum_{1,2} \frac{1}{Q^2} \left\{ u_{3a}^* \left[\frac{p_1 + p_3}{2m} - i\,\frac{\boldsymbol{\sigma}\times(p_1 - p_3)}{2m} \right]_{tr} u_{1a} \right\}$$

$$\times \left\{ u_{4a} \left[\frac{p_4 + p_2}{2m} - i\,\frac{\boldsymbol{\sigma}\times(p_2 - p_4)}{2m} \right]_{tr} u_{2a} \right\}$$

The first terms in each of the brackets represent currents due to motion of the electron transverse to Q and the second terms represent the transverse components of the magnetic dipole of each. So altogether it appears that the correction arises from current-current, current-dipole, and dipole-dipole interactions between the electrons. These interactions are expected even on the basis of classical theory and were described by Breit before quantum electrodynamics, hence are referred to as the Breit interaction.

Consider the dipole-dipole term arising in the correction factor. Since $Q = p_1 - p_3 = p_2 - p_4$ it is

$$\sum_{1,2} (\sigma_1 \times Q)_{tr}\,(\sigma_2 \times Q)_{tr}/Q^2$$

But since $\boldsymbol{\sigma}\times\mathbf{Q}$ is zero when $\boldsymbol{\sigma}$ and $\mathbf{Q}$ have the same direction, the sum could as well be over all three directions and then it is equivalent to a dot product. That is, this term of the correction is

$$(\sigma_1 \times Q)\cdot(\sigma_2 \times Q)/Q^2$$

By taking the Fourier transform † this will be seen to be the momentum representation of the interaction between two dipoles as was stated.

Note that the approximation $q_4 \sim (v/c)Q$ used above applies only between positive energy states. For, if one of the states represents a positron, then

$$q_4 = E_1 - E_2 \neq 0$$

$$= 2m$$

However, 2m is very large, so the correction is still small. It is necessary to redo the analysis nevertheless.

ELECTRON-POSITRON INTERACTION

It would appear that, since the electron and positron are distinguishable, the Pauli principle would not require the interchange diagram, leaving as the only one Fig. 26-2.

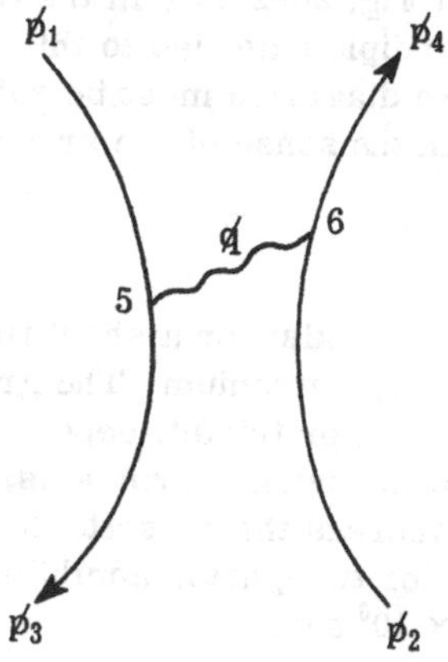

FIG. 26-2

But it is still possible by the same phenomenological reasoning to conceive of the diagram in Fig. 26-3, which would represent virtual annihilation of the electron and positron with the photon later creating a new pair. It turns out that it is necessary to regard an electron-positron pair as existing part of the time in the form of a virtual photon in order to obtain agreement with experiment.

---

† Notice that $(\sigma_1 \times Q) \cdot (\sigma_2 \times Q) \exp(-iQ \cdot x)$, which will appear in transform integral, is the same as $-(\sigma_1 \times \nabla) \cdot (\sigma_2 \times \nabla) \exp(-iQ \cdot x)$, where ∇ is the grad operator. This device enables an integration by parts, which greatly simplifies the process and the result. Thus, since the transform of $1/Q^2$ is $1/r$, the coupling is $-(\sigma_1 \times \nabla) \cdot (\sigma_2 \times \nabla)(1/r)$, which is the classical energy for interacting magnetic dipoles.

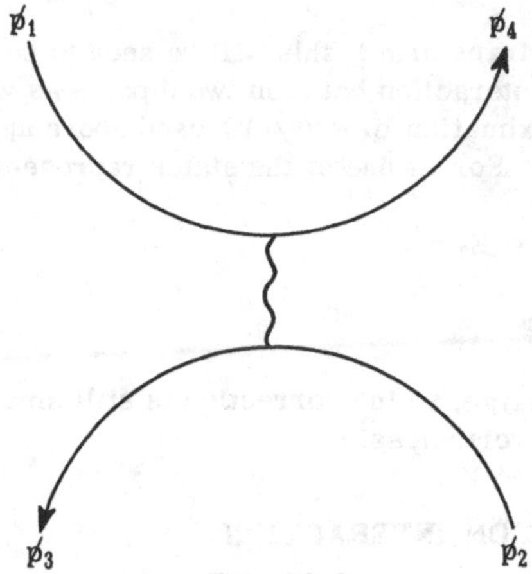

FIG. 26-3

From the point of view that positrons are electrons moving backward in time, Fig. 26-3 differs from Fig. 26-2 only in the interchange of the "final" states $\not{p}_3$, $\not{p}_4$. The Pauli principle extended to this case continues to operate; the amplitudes of the two diagrams must be subtracted, since they differ only in which outgoing (in the sense of the arrows) particle is which.

POSITRONIUM

An electron and positron can exist for a short time in a hydrogenlike bound state known as the atom positronium. The ground state of positronium is an S state and may be singlet or triplet, depending on the spin arrangement. As has been indicated in assigned problems, the ^1S state can annihilate only in two photons, whereas the ^3S state decays only by three-photon annihilation. The mean life for two-photon annihilation is $1/8 \times 10^9$ sec and for three photons it is $1/7 \times 10^6$ sec.

Problem: Check the mean life $1/8 \times 10^9$ sec for two-photon annihilation using the cross section already computed and using hydrogen wave functions with the reduced mass for positronium.

Figure 26-2 contributes the Coulomb potential holding the positronium together. The correction term (Breit's interaction) arising from this same diagram contributes a dipole-dipole or spin-spin interaction that is different in the ^3S and ^1S states (the current-current and spin-current interactions are the same for both states). Thus this amounts to a fine-structure separation of the ^3S and ^1S states which can be shown to be 4.8×10^{-4} ev.

In view of the fact that a photon has spin 1, and the ^1S state of positronium spin 0, conservation of angular momentum prohibits the process in Fig. 26-3 from occurring in the ^1S state. It does occur in the ^3S state, however. The term arising from this diagram is small and, therefore, constitutes another fine-structure splitting of the ^3S and ^1S levels. It can be shown to amount to 3.7×10^{-4} ev in the same direction as the spin-spin splitting. It is referred to as splitting due to the "new annihilation force."

In order to calculate the term arising from Fig. 26-3, one needs to compute

$$-(\tilde{u}_4\gamma_\mu u_1)(\tilde{u}_3\gamma_\mu u_2)/q^2$$

In this case $q^2 \approx 4m^2$ ($Q = 0$ in the C.G. system), and all matrix elements are 1 or 0 (regarding particles as essentially at rest in the positronium), so the result is just a number. This means that taking the Fourier transform one gets a δ function of the relative coordinate of the electron and positron for the interaction in real space. For this reason it is sometimes referred to as the "short-range" interaction of the electron and positron.

The combined fine-structure splitting due to the effects already outlined turns out to be represented by

$$(1/2)\,\alpha^2 \text{ Rydberg } (7/3)$$

where α is the fine-structure constant. This amounts to 2.044×10^5 Mc, using frequency as a measure of energy.

There is still another correction, however, not yet mentioned, arising from diagrams, such as Fig. 26-4, where the electron or positron may emit

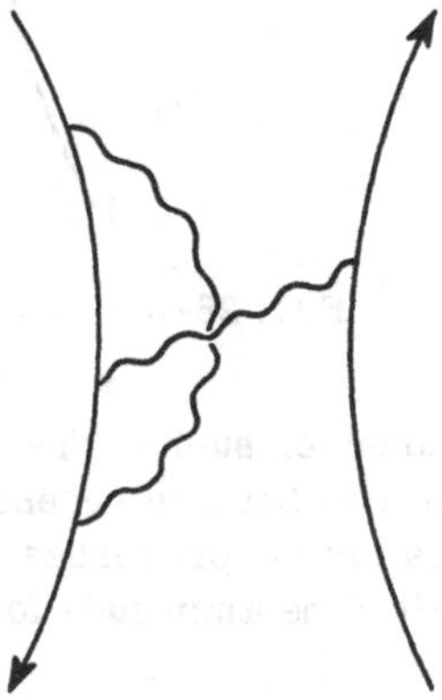

FIG. 26-4

and reabsorb its own photon. Taking this into account, the fine-structure splitting in positronium is given by †

$$(1/2)\,\alpha^2 \text{ Rydberg } \{(7/3) - [(32/9) + 2\ln 2](\alpha/\pi)\}$$

† Phys. Rev., 87, 848 (1952).

having a value of 2.0337×10^5 Mc. The experimental value for the positronium fine structure is 2.035 ± 0.003 Mc, so it is seen that this last correction, though of order α smaller than the main terms, is necessary to obtain agreement with experiment. It is referred to both in positronium and in hydrogen as the Lamb-shift correction because of its experimental observation by Lamb as the source of the small splitting between the ${}^2S_{1/2}$ and ${}^2P_{1/2}$ levels in hydrogen. In general, it comes under the heading of self-action of the electron, to be treated in more detail later.

TWO-PHOTON EXCHANGE BETWEEN ELECTRONS AND/OR POSITRONS

It is easy to imagine that processes, indicated by the diagrams in Fig. 26-5, may occur where two photons instead of one are exchanged. Although

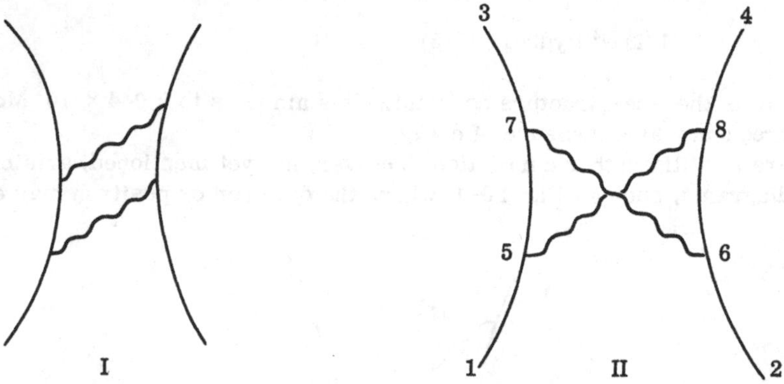

FIG. 26-5

it has not been necessary to consider such high-order processes to secure agreement with experiment, it may become necessary as experimental results improve. The amplitudes for the processes may be written down easily but their calculation is difficult. The amplitude for case II in space-time representation is, for example,

$$-e^4 \iiiint [K_+(3,7)\gamma_\nu\ K_+(7,5)\gamma_\mu K_+(5,1)]\,[K_+(4,8)\gamma_\mu K_+(8,6)\gamma_\nu \times K_+(6,2)]\ \delta_+(s^2_{7,6})\,\delta_+(s^2_{5,8})\ d\tau_5\, d\tau_6\, d\tau_7\, d\tau_8$$

or in momentum representation it is

$$-(4\pi)^2 e^4 \int \left(\tilde{u}_3 \gamma_\nu \frac{1}{\not{p}_1 - \not{k}_1 - m} \gamma_\mu u_1\right)\left(\tilde{u}_4 \gamma_\mu \frac{1}{\not{p}_2 - \not{k}_2 - m} \gamma_\nu u_2\right) \times \frac{d^4 \not{k}_1}{k_1^2\, k_2^2\, (2\pi)^4}$$

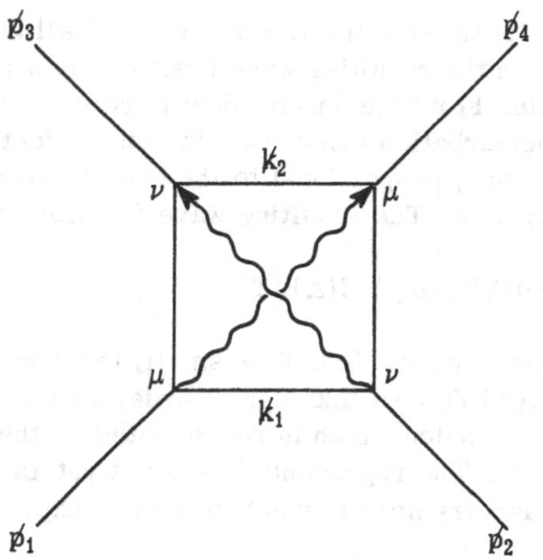

FIG. 26-6

where

$$\not{p}_2 - \not{k}_2 + \not{k}_1 = \not{p}_4 \qquad \text{or} \qquad \not{k}_2 = \not{p}_2 + \not{k}_1 - \not{p}_4$$

(see Fig. 26-6). Thus it is possible to determine $\not{k}_1$ and $\not{k}_2$ in terms of each other but not independently; that is, the momentum may be shared in any ratio between the two photons. It is for this reason that the integral over $\not{k}_1$ arises in the expression for the amplitude.

Twenty-seventh Lecture

SELF-ENERGY OF THE ELECTRON †

In Lecture 26 the following idea was introduced: An electron may emit and then absorb the same photon, as in Fig. 27-1. Then the propagation kernel for a free electron moving from point 1 to point 2 should include terms representing this possibility. Including only a first-order term (only one photon is emitted and absorbed), the resulting kernel is

$$K(2,1) = K_+(2,1) - ie^2 \iint K_+(2,4)\gamma_\mu K_+(4,3)\gamma_\mu K_+(3,1)\,\delta_+(s_{4,3}{}^2) \times d\tau_4\, d\tau_3 \tag{27-1}$$

The correction term in this equation is written down by an inspection of the diagram, following the usual procedure for scattering processes. In the present case, the initial and final momenta are identical. Therefore the

†R. P. Feynman, Phys. Rev., 76, 769 (1949); included in this volume.

nondiagonal elements in the perturbation matrix will all be zero. A diagonal element is one in which the resulting wave functions of a particle remain in the same eigenstate. For time-independent perturbations, it was shown in the development of perturbation theory that the only effect on such wave functions is a change in phase, proportional to the time interval T over which the perturbation is applied. The resulting wave function is

$$\exp(-iE_nT)\exp[-i(\Delta E)T] \tag{27-2}$$

Since the perturbation effect $(\Delta E)T$ is small, the second exponential can be expanded as $1 - i(\Delta E)T + \cdots$ and higher-order terms neglected. It is the second term of this expansion which is represented by the integral on the right side of Eq. (27-1). The representation is not yet an equality, since certain normalizing factors are different in the two expressions.

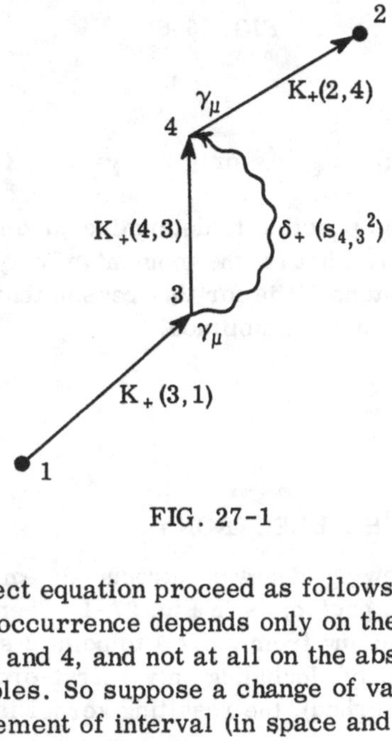

FIG. 27-1

To obtain the correct equation proceed as follows: First, it is clear that the probability of the occurrence depends only on the interval in space and time between points 3 and 4, and not at all on the absolute values of the space and time variables. So suppose a change of variable is made so that $d\tau_4$ represents the element of interval (in space and time) between 3 and 4. Then write the integral in Eq. (27-1)

$$\iint \tilde{f}(4)\gamma_\mu K_+(4,3)\gamma_\mu \delta_+(s_{4,3}^2) f(3)\, d\tau_4\, d^3x_3\, dt_3 \tag{27-3}$$

where it is clear that the operators K_+ and δ_+ depend only on the interval 3-4.

Second, expression (27-2) contains the time-dependent part of the wave function, $\exp(-iE_nt)$, because it was assumed that the wave functions used did not contain time factors. In Eq. (27-3), f(3), f(4) do already include the time-dependent part, so it should be omitted in Eq. (27-2).

Third, the normalization of wave functions is different for the two approaches. For the development that led to Eq. (27-2), the normalization

$$\int \Psi^* \Psi \, dv = 1$$

was used. For the present development the normalization is

$$\int u^* u \, dv = (2E/cm^3) \cdot V \tag{27-4}$$

Thus, to establish an equality, expression (27-3) must be divided by the normalizing integral of Eq. (27-4).

The resulting expression is

$$-i\,\Delta ET = \frac{-ie^2 \iint \tilde{f}(4)\,\gamma_\mu\, K_+(4,3)\,\gamma_\mu\, \delta_+(s_{4,3}{}^2) f(3)\, d\tau_4\, d^3x_3\, dt_3}{2E \cdot V}$$

The integral over d^3x_3 gives a V which cancels with the denominator, and the integral over dt_3 gives a T which cancels with the left-hand side, so finally

$$2E\,\Delta E = +e^2 \int \tilde{u}\,\gamma_\mu\, K_+(4,3)\,\gamma_\mu\, \delta_+(s_{4,3}{}^2) u \, d\tau_4 \tag{27-5}$$

Note that the integral is relativistically invariant. Further, since p is the same before and after the perturbation and $E^2 = m^2 + p^2$, the change in E can be taken as a change in the mass of the electron, from

$$2E\,\Delta E = 2m\,\Delta m$$

Using this expression, and transforming to momentum space,

$$\Delta m = \frac{4\pi e^2}{2mi} \int \tilde{u}\left(\gamma_\mu \frac{1}{\not{p} - \not{k} - m}\gamma_\mu\right) u \, \frac{d^4k}{(2\pi)^4}\frac{1}{k^2} \tag{27-6}$$

The integrand may be rewritten from

$$\gamma_\mu \frac{1}{\not{p} - \not{k} - m}\gamma_\mu = \frac{\gamma_\mu(\not{p} - \not{k} + m)\gamma_\mu}{\not{p}^2 - 2p \cdot k + k^2 - m^2} = \frac{2m + 2\not{k}}{k^2 - 2p \cdot k}$$

using $\not{p}u = mu$ and the relations of Lecture 10. Then Eq. (27-6) becomes

$$\Delta m = \frac{4\pi e^2}{i} \int \frac{2m + 2\not{k}}{k^2 - 2p \cdot k}\, \frac{d^4k}{(2\pi)^4}\frac{1}{k^2} \tag{27-6'}$$

This integral is divergent, and this fact presented a major obstacle to quantum electrodynamics for 20 years. Its solution requires a change in the fundamental laws. Thus suppose that the propagation kernel for a photon is

$(1/k^2)c(k^2)$ instead of just $(1/k^2)$, where $c(k^2)$ is so chosen that $c(0) = 1$ and $c(k^2) \to 0$ as $k^2 \to \infty$. In space representation the modification takes the form

$$\delta_+(s_{1,2}{}^2) \to f_+(s_{1,2}{}^2) = \int (1/k^2)c(k^2) \exp(-ik \cdot x)\, d^4k/(2\pi)^4 \tag{27-7}$$

The new function f_+ differs significantly from δ_+ only for small intervals. This is clear from the fact that if the high-frequency components are removed from the Fourier expansion of a function, only the short-range details are modified. In the present case the size of the interval over which the function is modified can be described roughly as follows: Consider a large number, λ^2, and suppose that so long as $k^2 \ll \lambda^2$, $c(k^2) \approx 1$. Then (from the exponential term) differences will occur when the interval $s^2 \approx 1/\lambda^2$. Call

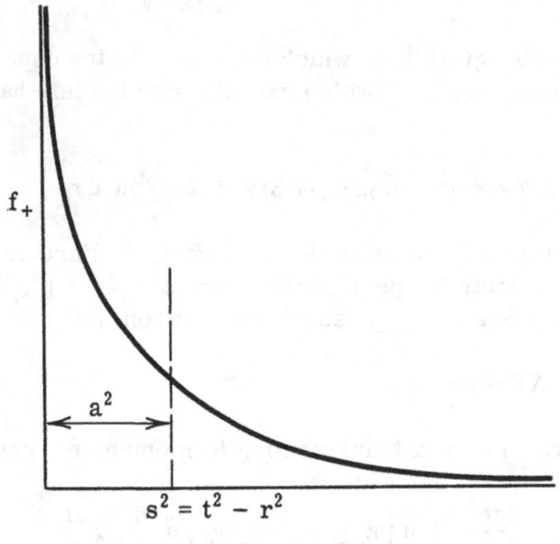

FIG. 27-2

this value a^2, and the general behavior of f_+ is shown by Fig. 27-2. Thus a^2 is sort of a "mean width" of f_+. If $a^2 \ll 1$, as assumed, then when

$$t^2 - r^2 = a^2 \qquad t - r \approx a^2/2r \tag{27-8}$$

which is the size of the interval. The significance of the form of $f_+(s^2)$ can be understood from the following. The original function, $\delta_+(s^2)$ differs from zero only when $s^2 = t^2 - r^2 = 0$. That is to say, an electromagnetic signal can reach a point at distance r only at a time t such that $t^2 - r^2 = 0$ or $t = r$ (i.e., the speed of light is 1). This is no longer true for $f_+(s^2)$. The departure is obtained by a measure of $t - r$. But, by Eq. (27-8), for all values of $r \gg a$ this measure is negligible. Thus, depending on λ^2, the laws will be found unaffected over any practical distance.

Choosing $\lambda^2 \gg m^2$, a practical (and general) representation of $c(k^2)$ is

$$c(k^2) = \int G(\lambda)\, d\lambda\, (-\lambda^2)(k^2 - \lambda^2)^{-1}$$

and the simple form is suggested,

$$c(k^2) = -\lambda^2/(k^2 - \lambda^2)$$

From this, obtain the propagation kernel as

$$1/k^2(-\lambda^2)(k^2 - \lambda^2)^{-1} = 1/k^2 - 1/(k^2 - \lambda^2)$$

The second term is that for the propagation of a photon of mass λ; however, the minus sign in front of the term has not been explained so far from this point of view.

A convenient representation for this kernel is the integral

$$-\int_0^{\lambda^2} dL/(k^2 - L)^2 \tag{27-9}$$

Introducing this kernel into Eq. (27-6′) in place of $1/k^2$ gives

$$\int \frac{2m + 2\not{k}}{k^2 - 2p \cdot k} \frac{d^4k}{k^2} \left(\frac{-\lambda^2}{k^2 - \lambda^2} \right) \tag{27-10}$$

which can be written as the sum of two integrals, which differ only by having m or $\not{k}$ in the numerator, that is, m or k_σ (since $\not{k} = k_\sigma \gamma_\sigma$).

METHOD OF INTEGRATION OF INTEGRALS APPEARING IN QUANTUM ELECTRODYNAMICS

We shall need to do many integrals of a form similar to the preceding one. A method has been worked out to do these fairly efficiently. We now stop to describe this method of integration.

Everything will be based on the following two integrals:†

$$\int_{-\infty}^{\infty} \frac{(1; k_\sigma)\, d^4k}{(2\pi)^4 (k^2 + i\epsilon - L)^3} = (32\pi^2 iL)^{-1}\, (1;0) \tag{27-11}$$

$$\int_0^1 [ax + b(1-x)]^{-2}\, dx = 1/ab \tag{27-12}$$

In Eq. (27-11), to write a little more compactly, we use the notation $(1;k_\sigma)$ to mean that either 1 or k_σ is in the numerator, in which case, on the right-hand side the (1;0) is 1 or 0, respectively. To prove the first of these, note

†R. P. Feynman, Phys. Rev., **76**, 769 (1949); included in this volume. Note that in the article d^4k is equivalent to $4\pi^2[d^4k/(2\pi)^4]$ in our notation.

that, if k_σ is in the numerator, the integrand is an odd function. Thus the integral is zero. With 1 in the numerator, contour integration is employed. Write the integral

$$\int\int\int_{-\infty}^{\infty}\int [\omega^2 + i\epsilon - (L + k^2)]^{-3}\, d\omega\, d^3\mathbf{k}$$

Then for $\epsilon \ll L + k^2$, there are poles at $\omega = \pm[(L + k^2)^{1/2} - i\epsilon]$, and contour integration of ω gives

$$\int_{-\infty}^{\infty} [\omega^2 + i\epsilon - (L + k^2)]^{-1}\, d\omega = 2\pi i[-2(L + k^2)^{-1/2}]^{-1}$$

with the contour in the upper half-plane. Two differentiations with respect to L give

$$\int_{-\infty}^{\infty}[\omega^2 + ie - (L+k^2)]^{-3}\, d\omega = (6\pi/16i)(L + k^2)^{-5/2}$$

Then the remaining integral is

$$\int\int_{-\infty}^{\infty}\int (L+k^2)^{-5/2}\, d^3\mathbf{k} = 4\pi \int_0^{\infty}(L+k^2)^{-5/2}\, k^2\, dk$$

$$= 4\pi\, [k^3/3L(L+k^2)^{3/2}]\Big|_0^{\infty} = 4\pi/3L$$

which proves Eq. (27-11). If $k - p$ is substituted for the variable of integration in Eq. (27-11), the result is

$$\int_{-\infty}^{\infty} \frac{(1;k_\sigma)\, d^4k}{(2\pi)^4(k^2 - 2p\cdot k - \Delta)^3} = [32\pi^2 i(p^2 + \Delta)]^{-1}(1;p_\sigma) \tag{27-13}$$

By differentiating both sides of Eq. (27-13) with respect to Δ or with respect to p_j, there follows directly

$$\int_{-\infty}^{\infty} \frac{(1;k_\sigma;k_\sigma k_j)d^4k}{(2\pi)^4(k^2 - 2p\cdot k - \Delta)^4} = -\frac{[1;p_\sigma;p_\sigma p_j - (1/2)\delta_{\sigma j}(p^2 + \Delta)]}{96\pi^2 i(p^2 + \Delta)^2}$$

Further differentiations give directly successive integrals including more k factors in the numerator and higher powers of $(k^2 - 2p\cdot k - \Delta)$ in the denominator.

Twenty-eighth Lecture

SELF-ENERGY INTEGRAL WITH AN EXTERNAL POTENTIAL

Last time it was found that the self-energy of the electron is equivalent to a change in mass

$$\Delta m = \frac{4\pi e^2}{2mi} \int \frac{\tilde{u}(2m + 2\not{k})\, u}{k^2 - 2p \cdot k} \left(\frac{-\lambda^2}{k^2 - \lambda^2}\right) \frac{1}{k^2} \frac{d^4 k}{(2\pi)^4} \tag{28-1}$$

and that this could also be expressed in terms of integrals,

$$I = -\int_0^{\lambda^2} dL \int \frac{(1;k_\sigma)}{(k^2 - 2p \cdot k)(k^2 - L)^2} \frac{d^4 k}{(2\pi)^4} \tag{28-2}$$

It was also found that

$$\int \frac{(1;k_\sigma)}{(k^2 - 2p \cdot k - \Delta)^3} = (32\pi^2 i)^{-1} (p^2 + \Delta)^{-1} \tag{28-3}$$

Using the definite integral

$$\frac{1}{ab^2} = \int_0^1 \frac{2(1-x)\, dx}{[ax + b(1-x)]^3} \tag{28-4}$$

the denominator of the integrand of Eq. (28-2) may be expressed as

$$\frac{1}{(k^2 - 2p \cdot k)(k^2 - L)^2} = \int_0^1 \frac{2(1-x)\, dx}{[k^2 - 2xp \cdot k - L(1-x)]^3}$$

so that Eq. (28-2) becomes

$$I = -\int_0^{\lambda^2} dL \int\int_0^1 \frac{d^4 k\, (1;k_\sigma)\, 2(1-x)\, dx}{[k^2 - 2xp \cdot k - L(1-x)]^3 (2\pi)^4} \tag{28-5}$$

The integral over k can be done by using Eq. (28-3) with the substitutions xp for p and L(1 − x) for Δ, giving

$$I = -\int_0^{\lambda^2} dL \int_0^1 \frac{(1;p_\sigma)\, 2(1-x)\, dx}{[32\pi^2 i]\,[x^2 p^2 + L(1-x)]} \qquad p^2 = m^2$$

The integral over L is elementary and gives

$$I = -2(32\pi^2 i)^{-1} \int_0^1 dx\, (1;xp_\sigma) \ln [(1-x)\lambda^2 + m^2 x^2 / m^2 x^2]$$

When $\lambda^2 \gg m^2$, it is legitimate to neglect $m^2 x^2$ in the numerator [it is true that when $x \approx 1$, $(1 - x)\lambda^2$ is *not* much larger than $m^2 x^2$, but the interval over which this is true is so small, for $\lambda^2 \gg m^2$, that the error is small], so that, when the x integration is performed,†

---

† $\int_0^1 \ln [x^{-2}(1-x)]\, dx = 1 \qquad \int_0^1 x \ln [x^{-2}(1-x)]\, dx = -1/4$

$$I \approx -(32\pi^2 i)^{-1}\{2[\ln(\lambda^2/m^2) + 2]; p_\sigma[\ln(\lambda^2/m^2) - 1/2]\}$$

$$\lambda >> m^2$$

The change in mass is [from Eq. (28-1)]

$$\Delta m = (4\pi^2/2mi)(-32\pi^2 i)^{-1}(\tilde{u}\{2m[2\ln(\lambda^2/m^2) + 2]$$
$$+ 2\not{p}[\ln(\lambda^2/m^2) - (1/2)]\}u)$$

Since $\not{p}u = mu$ and $(\tilde{u}u) = 2m$, this can be simplified to

$$\Delta m/m = (e^2/2\pi)[3\ln(\lambda/m) + (3/4)] \qquad (28\text{-}6)$$

Now $(e^2/2\pi)$ is about 10^{-3}, so that even if λ is many times m, the fraction change in mass will not be large. The interpretation of this result is as follows. There is a shift in mass which depends on λ and hence cannot be determined theoretically. One can imagine an experimental mass and a theoretical mass which are related by

$$m_{exp} = m_{th} + \Delta m \qquad (28\text{-}7)$$

All our measurements are of m_{exp}, that is, self-action is included, and m_{th}, the mass without self-action, cannot be determined. More accurately stated,

$$\left\{\begin{matrix}\text{A theory using } m_{th} \text{ and}\\ e^2/\hbar c \text{ self-action}\end{matrix}\right\} \text{ is equivalent to } \left\{\begin{matrix}\text{a theory using } m_{exp}\text{, } \textit{plus}\\ e^2/\hbar c \text{ self-action } \textit{minus}\\ \Delta m \text{ as computed for a}\\ \text{free particle}\end{matrix}\right\}$$

When the electron is free, the $e^2/\hbar c$ self-action term exactly cancels the Δm term and a theory using m_{exp} is exactly correct. When the electron is not free, $e^2/\hbar c$ self-action is not quite equal to the Δm term and there is a small correction to a theory using m_{exp}. This effect leads to the Lamb shift in the hydrogen atom, and, in order to calculate such effects, we shall now consider the effect of self-action on the scattering of an electron by an external potential.

SCATTERING IN AN EXTERNAL POTENTIAL

The diagram for scattering in an external potential is shown in Fig. 28-1, and the relationships for this process, excluding the possibility of self-action, are as follows:

Potential: $\not{a}(q) = \gamma_t(4\pi Ze/Q^2)\delta(q_4)$ for Coulomb potential
Matrix element: $M = -ie(\tilde{u}_2\not{a}u_1)$
Conservation relation: $\not{p}_2 = \not{p}_1 + \not{q}$

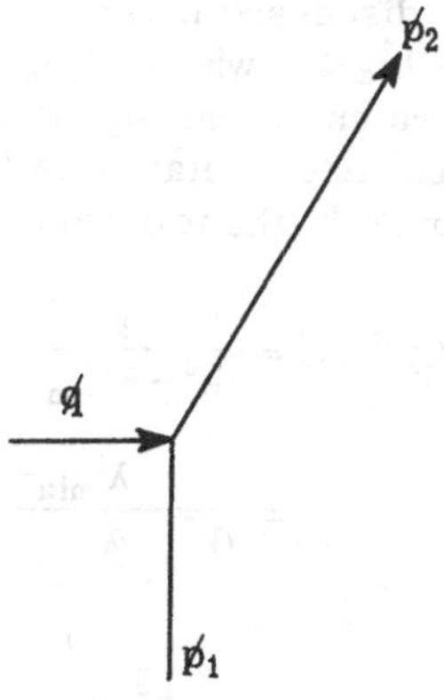

FIG. 28-1

First-order self-action will produce the diagrams shown in Fig. 28-2. The amplitude for process is obtained in the usual manner. For example, diagram I gives

$$I_1 = \frac{4\pi e^2}{i}\int\left(\tilde{u}_2\gamma_\mu \frac{1}{\not p_2 - \not k - m} \not a \frac{1}{p_1 - \not k - m}\gamma_\mu u_1\right)(k^2)^{-1}(2\pi)^{-4}\, d^4k$$

Rationalizing the denominators and inserting the convergence factor, this becomes

$$I_1 = \frac{4\pi e^2}{i}\int \frac{(\tilde{u}_2\gamma_\mu[\not p_2 - \not k + m]\not a[\not p_1 - \not k + m]\gamma_\mu u_1)}{(k^2 - 2p_2\cdot k)(k^2 - 2p_1\cdot k)} \quad \frac{-\lambda^2}{k^2 - \lambda^2}$$

$$\times (2\pi)^{-4}\frac{d^4k}{k^2} \tag{28-8}$$

This expression also happens to diverge for small photon momenta (k) (a result which has been called the ''infrared catastrophe,'' but which has a

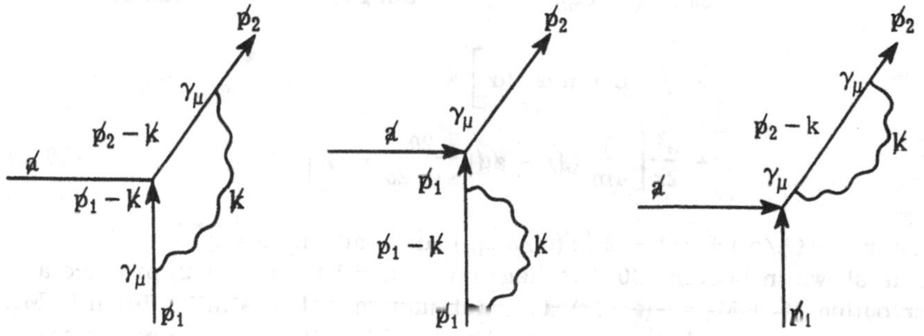

FIG. 28-2

clear physical interpretation, discussed later). Temporarily the k^2 under d^4k will be replaced by $(k^2 - \lambda^2_{min})$, where $\lambda^2_{min} \ll m^2$, to make the integral convergent. This is equivalent to cutting off the integral somewhere near $k = \lambda_{min}$, and the physical interpretation is left to Lectures 29 and 30. To facilitate the integration over k, the following identity is used:

$$-\int_{\lambda^2_{min}}^{\lambda^2} (k^2 - L)^{-2}\, dL = \frac{1}{k^2 - \lambda^2_{min}} - \frac{1}{k^2 - \lambda^2}$$

$$= \frac{\lambda^2_{min} - \lambda^2}{(k^2 - \lambda^2_{min})(k^2 - \lambda^2)} \approx \frac{-\lambda^2}{k^2 - \lambda^2} \times \frac{1}{k^2 - \lambda^2_{min}}$$

since $\lambda^2 \gg m^2 \gg \lambda^2_{min}$. This substitution produces integrals of the form

$$-\int_{\lambda^2_{min}}^{\lambda^2} dL \int \frac{(1;k_\sigma;k_\sigma k_\tau)(2\pi)^{-4}\, d^4k}{(k^2 - 2p_1 \cdot k)(k^2 - 2p_2 \cdot k)(k^2 - L)^2}$$

To evaluate these integrals, we make use of the identity

$$(ab)^{-1} = \int_0^1 dy/[ay + b(1-y)]^2$$

so that

$$\frac{1}{(k^2 - 2p_1 \cdot k)(k^2 - 2p_2 \cdot k)} = \int_0^1 \frac{dy}{(k^2 - 2\not{p}_y \cdot k)^2}$$

where $\not{p}_y = y\not{p}_1 + (1-y)\not{p}_2$. Performing integrations in the order, k, L, y, and using the appropriate integrals in Eq. (28-6) gives as the matrix to be taken between states u_2 and u_1

$$M_1 = \frac{e^2}{2\pi}\left[2\left(\ln\frac{m}{\lambda_{min}} - 1\right)\left(1 - \frac{2\theta}{\tan 2\theta}\right) + \theta \tan\theta + \frac{4}{\tan 2\theta} \times \int_0^\theta \alpha \tan\alpha\, d\alpha\right]\not{a}$$

$$+ \frac{e^2}{2\pi}\left[\frac{1}{4m}(\not{q}\not{a} - \not{a}\not{q})\frac{2\theta}{\sin 2\theta} + r\not{a}\right] \qquad (28\text{-}9)$$

where $r = \ln(\lambda/m) + 9/4 - 2\ln(m/\lambda_{min})$ and $4m^2 \sin^2\theta = q^2$.

It is shown in Lecture 30 that diagrams II and III (Fig. 28-2) produce a contribution $M_2 + M_3 = -(e^2/2\pi)r\not{a}$, which just cancels a similar term in M_3. When q is small, $\theta \approx (q^2)^{1/2}/2m$, and the sum $M_1 + M_2 + M_3$ can be approximated by

$$M \approx \frac{e^2}{4\pi}\left[\frac{1}{2m}(\not{q}\not{a} - \not{a}\not{q}) + \frac{4q^2}{3m^2}\not{a}\left(\ln\frac{m}{\lambda_{min}} - \frac{3}{8}\right)\right] \tag{28-10}$$

The $(\not{q}\not{a} - \not{a}\not{q})$ can be written out

$$(\not{q}\not{a} - \not{a}\not{q}) = \gamma_\mu\gamma_\nu(q_\mu a_\nu - a_\mu q_\nu)$$

But q_μ is the gradient operator so this can be written, in coordinate representation,

$$\gamma_\mu\gamma_\nu(\nabla_\mu A_\nu - \nabla_\nu A_\mu) = +\gamma_\mu\gamma_\nu F_{\mu\nu}$$

[see Eq. (7-1)]. Reference to page 54 shows that the effect of a particle's having an anomalous magnetic moment is to subtract a potential $\mu\gamma_\mu\gamma_\nu F_{\mu\nu}$ from the ordinary potential $\not{a} = \gamma_\mu A_\mu$ appearing in the Dirac equation. Since this is precisely what the first term of Eq. (28-10) does, one can say that this part of the self-action correction looks like a correction to the electron's magnetic moment, so that

$$\mu_{elec} = (e/2m)[1 + (e^2/2\pi)]$$

Note that this result [and (28-9) and (28-10)] does not depend on the cutoff λ, and hence λ can now be taken to be infinity.†

Twenty-ninth Lecture

It has been shown that when a particle is scattered by a potential, the primary effect is that of $\not{a}$, and that for diagram I (Fig. 28-2) a correction term arises which is

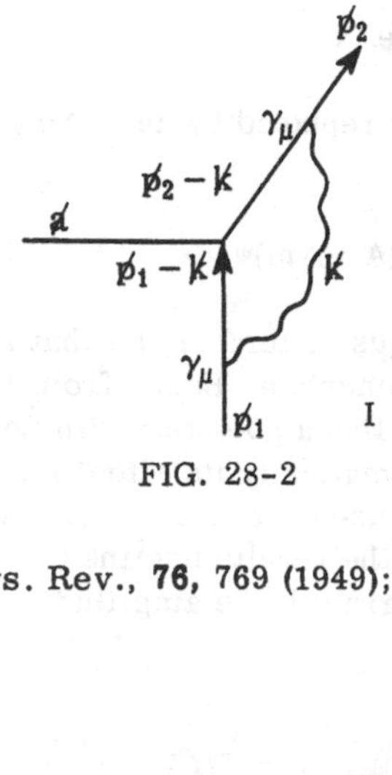

FIG. 28-2

† R. P. Feynman, Phys. Rev., **76**, 769 (1949); included in this volume.

$$\frac{e^2}{2\pi}\left[2\left(\ln\frac{m}{\lambda_{min}} - 1\right)\left(1 - \frac{2\theta}{\tan 2\theta}\right) + \theta \tan\theta + \frac{4}{\tan 2\theta}\right.$$

$$\left.\times \int_0^\theta \alpha \tan\alpha \right] a\!\!/ + \frac{e^2}{8\pi m}(q\!\!/ a\!\!/ - a\!\!/ q\!\!/)\frac{2\theta}{\sin 2\theta} + \frac{e^2}{2\pi} r a\!\!/$$

It remains to show that the combined effect of diagrams II and III (Fig. 28-2),

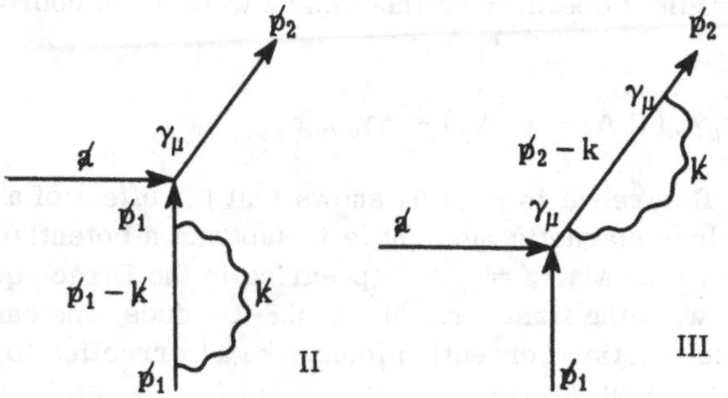

FIG. 28-2

when considered along with the effect of the mass correction, is another correction term,

$$-(e^2/2\pi)r a\!\!/$$

just canceling the last term in the preceding expression. It is recalled that the necessity for considering the effect of the mass correction together with the self-action represented in diagrams I, II, and III is that the theory being developed must contain the experimental mass rather than the "theoretical" mass.

Suppose that in the Dirac equation

$$(i\nabla\!\!\!/ - m_{th})\Psi = eA\!\!\!/\Psi$$

m_{th} , the theoretical mass, is replaced by $m - \Delta m$, where m is the experimental mass; then

$$(i\nabla\!\!\!/ - m)\Psi = e(A\!\!\!/ + \Delta m)\Psi$$

The mass correction Δm is just a number, so that in momentum representation it is a δ function of momentum. Hence from the form of the foregoing equation, it is seen to behave like a potential with zero momentum and involves no matrices. Diagrammatically its effect may be represented as in Fig. 29-1. The minus sign is used because the effect of the mass correction Δm is to be subtracted from the results obtained from diagrams I, II, and III (Fig. 28-2) alone. For diagram II the amplitude would appear to be

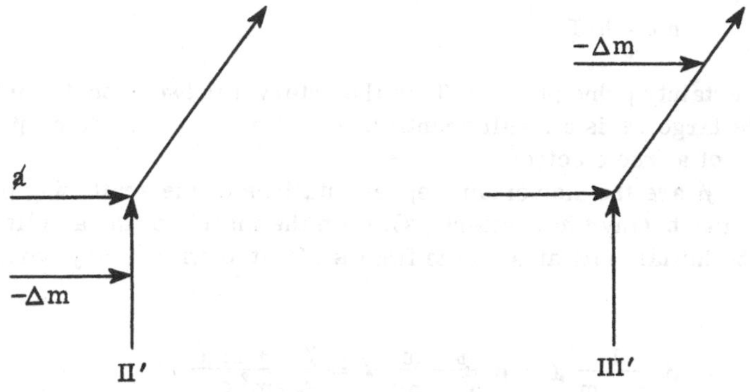

FIG. 29-1

$$\tilde{u}_2 \not{a} \frac{1}{\not{p}_1 - m}\left(\frac{4\pi e^2}{i} \int \gamma_\mu \frac{1}{\not{p}_1 - \not{k} - m} \gamma_\mu \frac{1}{k^2 - \lambda_{min}} \frac{d^4k}{(2\pi)^4} \frac{-\lambda}{k^2 - \lambda^2}\right)$$

and for diagram II′ (Fig. 29-1),

$$-\tilde{u}_2 \not{a} [1/(\not{p}_1 - m)] (\Delta m) u_1$$

But the part of the amplitude for diagram II (Fig 28-2) contained in the parentheses is just $\Delta m u_1$, so that II and II′ seem to cancel. A similar result applied for diagrams III and III′. This is an error, however, arising from the fact that both of these amplitudes are infinite, owing to the factor $\not{p} - m$ in the denominator. Hence their difference is indeterminate. But by subtracting them properly it will be found that their difference does not vanish.

The method proposed to accomplish this subtraction will, in fact, give the combined effect of the self-action and mass correction of both diagrams II and III and II′ and III′. It is based on the fact that an electron is never actually free. An electron's history will have always involved a series of scatterings, as will its future. These scatterings will be considered as occurring at long but finite time intervals. It will be sufficient to calculate the effect of self-action and the mass correction between any two of these scatterings, since the result will evidently be the same between each pair of them. Then, the effect will be accounted for simply by regarding a correction, equal to that calculated for one of the intervals between scatterings, as being associated with the potential at each scattering (number of intervals equals number of scatterings). Then, considering a single scattering event as here, this correction to the potential represents all the effects of diagrams II, III, II′, and III′.

For an electron which is not quite free, $p^2 \neq m^2$ exactly, but instead

$$p^2 = m^2(1 + \epsilon)^2$$

where

$$m\epsilon = \hbar/T$$

by the uncertainty principle, and T is the interval between scatterings. Since T is large, ϵ is a small quantity. Let $\not{p} = (1+\epsilon)\not{p}_0$, where $\not{p}_0$ is the momentum of a free electron.

If $\not{a}$ and $\not{b}$ are the momentum representatives of the scattering potentials at a and b (any two scatterings), then the matrix of the amplitude to go from the initial state at a to the final state at b without any perturbations is

$$\not{b}\,\frac{1}{\not{p}-m}\,\not{a} = \not{b}\,\frac{\not{p}+m}{p^2-m^2}\,\not{a} = \frac{\not{b}(\not{p}+m)\not{a}}{2m^2\epsilon}$$

up to terms of order ϵ. With the perturbations of self-action and mass correction, this matrix is

$$i4\pi e^2 \int \not{b}\,\frac{1}{\not{p}-m}\,\gamma_\mu\,\frac{1}{\not{p}-\not{k}-m}\,\gamma_\mu\,\frac{1}{\not{p}-m}\,\not{a}\,\frac{d^4k}{k^2-\lambda^2_{min}}\,\frac{-\lambda^2}{k^2-\lambda^2}$$

$$-\not{b}\,\frac{1}{\not{p}-m}\,\Delta m\,\frac{1}{\not{p}-m}\,\not{a}$$

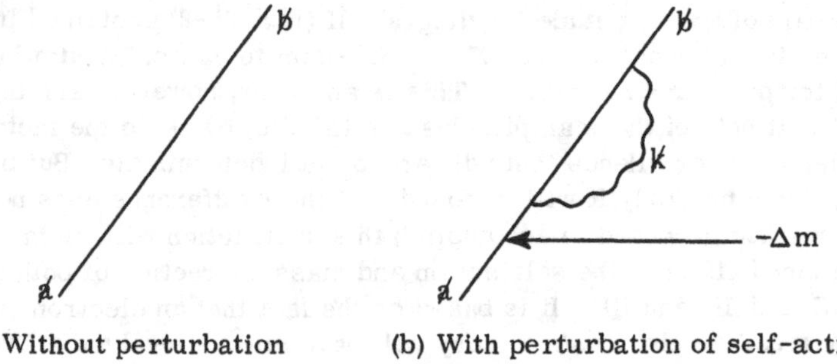

(a) Without perturbation (b) With perturbation of self-action and mass correction

FIG. 29-2

It is the value of this matrix compared to that of the unperturbed matrix which gives the desired correction term (see Fig. 29-2).

Problem: Show that for two noncommuting (or commuting) operators A and B, the following expansion is true:

$$\frac{1}{A+B} = \frac{1}{A} - \frac{1}{A}B\frac{1}{A} + \frac{1}{A}B\frac{1}{A}B\frac{1}{A} + \cdots$$

Using the result of the preceding problem, one can write

$$\frac{1}{\not p - \not k - m} = \frac{1}{\not p_0 + \epsilon \not p_0 - \not k - m} \approx \frac{1}{\not p_0 - \not k - m} - \frac{1}{\not p_0 - \not k - m} \times \epsilon \not p_0 \frac{1}{\not p_0 - \not k - m} + \cdots$$

so that the foregoing matrix becomes

$$i4\pi e^2 \int \not p \frac{\not p + m}{2m^2\epsilon} \gamma_\mu \frac{1}{\not p_0 - k - m} \gamma_\mu \frac{\not p + m}{2m^2\epsilon} \not a \frac{d^4k}{k^2 - \lambda^2_{min}} \frac{-\lambda^2}{k^2 - \lambda^2}$$

$$-i4\pi e^2 \int \not p \frac{\not p + m}{2m^2\epsilon} \gamma_\mu \frac{1}{\not p_0 - \not k - m} p_0 \frac{1}{\not p_0 - \not k - m} \gamma_\mu \frac{d^4k}{k^2 - \lambda^2_{min}}$$

$$\times \frac{\not p + m}{2m^2} \not a \left(\frac{-\lambda^2}{k^2 - \lambda^2}\right) - \not p \frac{\not p + m}{2m^2\epsilon} \Delta m \frac{\not p + m}{2m^2\epsilon} \not a$$

The first and last terms are identical, up to terms of order ϵ, hence may be canceled. The integral in the second term has already been done essentially in computing diagram I (Fig. 28-2), except here $\not p_0$ replaces $\not a$, $\not p_1$, and $\not p_2$, so that $\not q = \not p_2' - \not p_1 = 0$ in this case and gives the result

$$-\frac{e^2}{2\pi} r \not p \frac{\not p + m}{2m^2\epsilon} \not p_0 \frac{\not p + m}{2m^2} \not a$$

To this order in ϵ the $\not p$'s in the numerator may be replaced by $\not p_0$'s. It is also noted that since $\not p_0 u = mu$,

$$(\not p_0 + m) \not p_0 (\not p + m) \equiv 2m^2(\not p + m)$$

so that the foregoing result may be written

$$-\frac{e^2}{2\pi} r \not p \frac{\not p + m}{2m^2 \varepsilon} \not a$$

This is just $-(e^2/2\pi)r$ times the matrix for no perturbation. Hence the *correction term* due to diagrams II, III, II′, and III′ is obtained simply by replacing the scattering potential $\not a$ by $-(e^2/2\pi)r\not a$, as was stated earlier.

It should be noted that the difficulty in obtaining the proper subtraction of the self-action and mass corrections just clarified does not represent a ''divergence'' problem of quantum electrodynamics. It is a typical problem which could as well arise in nonrelativistic quantum mechanics if, for example, one chose some nonzero value as a reference of potential, that is, regarded a free electron as moving in a uniform nonzero potential. It may be easily verified that this would give rise to an ''energy correction'' for the free electron analogous to the mass correction involved here. Then in computing the amplitude for a scattering process where one used a "theoretical energy" and subtracted the effect of the "energy correction," the difference of infinite terms would appear if one used free-electron wave functions. In this simple case the infinite term would, indeed, cancel upon proper subtraction but in principle the problem is the same as the present one.

Finally, the complete correction term arising from self-action and mass correction is

$$\frac{e^2}{2\pi}\left[2\left(\ln\frac{m}{\lambda_{min}} - 1\right)\left(1 - \frac{2\theta}{\tan 2\theta}\right) + \theta\tan\theta + \frac{4}{\tan 2\theta}\right.$$

$$\left.\times \int_0^\theta \alpha\tan\alpha\, d\alpha\right] a\!\!\!/ + \frac{e^2}{8\pi m}(q\!\!\!/ a\!\!\!/ - a\!\!\!/ q\!\!\!/)\frac{2\theta}{\sin 2\theta}$$

RESOLUTION OF THE FICTITIOUS "INFRARED CATASTROPHE"

From the correction term just determined, it is seen that, to order e^2, the cross section for scattering of an electron with the emission of *no* photons is

$$\sigma = \sigma_0\left\{1 - \frac{e^2}{2\pi}\left[2\left(\ln\frac{m}{\lambda_{min}} - 1\right)\left(1 - \frac{2\theta}{\tan 2\theta}\right) + \begin{pmatrix}\text{terms not}\\ \text{dependent}\\ \text{on } \lambda_{min}\end{pmatrix}\right]\right\}$$

where σ_0 is the cross section for the potential $a\!\!\!/$ only. This cross section diverges logarithmically as $\lambda_{min} \rightarrow 0$, and it is this divergence which was formerly referred to as the "infrared catastrophe."

This result, however, arises from the physical fact that it is impossible to scatter an electron with the emission of *no* photons. When the electron is scattered, the electromagnetic field must change from that of a charge moving with momentum p_1 to that for momentum p_2. This change of the field is necessarily accompanied by radiation.

In the theory of brehmsstrahlung, it was shown that the cross section for emission of one low-energy photon is

$$\sigma = \sigma_0\frac{e^2}{\pi}\frac{d\Omega_\omega}{4\pi}\left(\frac{\omega p_1\, e}{p_1\cdot q} - \frac{\omega p_2\cdot e}{p_2\cdot q}\right)^2\frac{d\omega}{\omega}$$

Problem: Show that the integral over all directions and the sum over polarizations of the foregoing cross section is

$$\sigma = \sigma_0(2e^2/\pi)[1 - (2\theta/\tan 2\theta)]\, d\omega/\omega$$

where $\sin^2\theta = -(p\!\!\!/_2 - p\!\!\!/_1)^2/4m^2$. Thus the probability of emitting *any* photon between $k = 0$ and $k = K_m$ is

$$\sigma_0 \frac{2e^2}{\pi}\left(1 - \frac{2\theta}{\tan 2\theta}\right)\int_0^{K_m} \frac{d\omega}{\omega} = \sigma_0 \frac{2e^2}{\pi}\left(1 - \frac{2\theta}{\tan 2\theta}\right) \ln \frac{K_m}{\lambda_{min}}$$

which diverges logarithmically.

Therefore, the dilemma of the diverging scattering cross section actually arises from asking an improper question: What is the chance of scattering with the emission of no photons? Instead, one should ask: What is the chance of scattering with the emission of no photon of energy greater than K_m? For there will always be some very soft photons emitted.

Then, effectively, what is sought in answer to the last question is the chance of scattering and emitting no photon, the chance of emitting one photon of energy below K_m, and the chance of two and more photons below K_m (but these terms are of order e^4 and higher and hence are neglected).

Each of these terms is infinite, actually, but is kept finite temporarily by the artifice of the λ_{min}. Their sum, however, does not diverge, as may be seen by gathering the previous results and by writing

$$\begin{aligned} &\text{Chance of scattering and emitting no photon of energy} > K_m \\ &= \sigma_0 \left\{1 - \frac{e^2}{\pi}\left[2\left(\ln \frac{m}{\lambda_{min}} - 1\right)\left(1 - \frac{2\theta}{\tan 2\theta}\right) + \text{(terms independent of } \lambda_{min})\right]\right\} + \sigma_0 \frac{2e^2}{\pi}\left(1 - \frac{2\theta}{\tan 2\theta}\right) \ln \frac{K_m}{\lambda_{min}} + \text{(terms of order } e^4) \\ &= \sigma_0\left[\left(1 - \frac{e^2}{\pi} 2 \ln \frac{m}{K_m}\right)\left(1 - \frac{2\theta}{\tan 2\theta}\right)\right] + \left(\begin{matrix}\text{terms independent} \\ \text{of } \lambda_{min} \text{ and of} \\ \text{order } e^4\end{matrix}\right) \end{aligned}$$

This does not depend on λ_{min} and hence resolves the "infrared catastrophe." It has been shown by Bloch and Nordsieck that the same idea applies to all orders.†

It is interesting that the largest term in the quantum-electrodynamic corrections to the scattering cross section, namely,

$$-(2e^2/\pi)\,[1 - (2\theta/\tan 2\theta)]\,\ln (m/K_m)$$

may be obtained from classical electrodynamics, since such long wavelengths are involved. The other terms have small effects. To date, the scattering experiments have been accurate enough to verify the existence of the large term but not accurate enough to verify the exact contributions of the smaller terms. Hence they do not provide a nontrivial test of quantum electrodynamics.

These same considerations apply in any process involving the deflection

---

† F. Bloch and A. Nordsieck, Phys. Rev., **52**, 54 (1937).

of free electrons. The best way to handle the problem is to calculate everything in terms of the λ_{min} and then to ask only questions which can have a sensible answer as verified by the eventual elimination of the λ_{min}.

Problem: Prepare diagrams and integrals needed for the radiative corrections (of order e^2) to the Klein-Nishina formula. Do as much as possible and compare results with those of L. Brown and R. P. Feynman.†

Thirtieth Lecture

ANOTHER APPROACH TO THE INFRARED DIFFICULTY

Instead of introducing an artificial mass, assume no weak photons contribute. Thus we must subtract from the previous results the contributions of all photons with momentum magnitude less than some number $k_0 \gg \lambda$. The previous result is

$$\not{a}\{1 + (e^2/2\pi)[2 \ln (m/\lambda_{min} - 1)(1 - 2\theta/\tan 2\theta)] + \theta \tan \theta + (4/\tan 2\theta) \int_0^\theta y \tan y \, dy]\} \tag{30-1}$$

The term to be subtracted is

$$(e^2/2\pi) \int_0^{k_0} \gamma_\mu (p_2 - \not{k} + m)(k^2 - 2p_2 \cdot k_2)^{-1} \not{a}(\not{p}_1 - \not{k} + m) \times (k^2 - 2p_1 \cdot k_1)^{-1} \gamma_\mu \, d^4k/(k^2 - \lambda_{min}^2) \tag{30-2}$$

We assume $k_0 \ll p_1$ or p_2, and neglect both $\not{k}$ and the first two k^2 in this integral. Then using $\not{p}_1\gamma_\mu = 2p_\mu - \gamma_\mu \not{p}_1$, the integral is approximately

$$x = -\frac{e^2}{2\pi} \frac{\not{a}}{2} \int \left[\frac{p_{2\mu}}{p_2 \cdot k} - \frac{p_{1\mu}}{p_1 \cdot k}\right]^2 \frac{d^4k}{k^2 - \lambda_{min}^2} \tag{30-3}$$

Then

$$x = e^2/2\pi\{[1 - (2\theta/\tan 2\theta)][2 \ln (2k_0/\lambda_{min} - 1)] + [4\theta/\tan 2\theta] \times [(1/2\theta) \int_0^{2\theta} (y/\tan y) \, dy - 1]\} \tag{30-4}$$

This is the term to be subtracted from expression (30-1).

Using $\sin^2 \theta = q^2/4m^2$, for small q, Eq. (30-4) becomes

$$x = (e^2/2\pi)(2q^2/3m^2)[\ln (2k_0/\lambda_{min}) - (5/6)]$$

---

† Phys. Rev., **85**, 231 (1952).

Subtracting this from Eq. (30-1), also with q small, gives

$$\not{a}\,\{1 + (e^2/4\pi)(4q^2/3m^2)[\ln M/\lambda_{min}) - (3/8) - \ln(2k_0/\lambda_{min}) + (5/6)]\} \tag{30-5}$$

The last term is $[\ln (M/2k_0) + (11/24)]$.

EFFECT ON AN ATOMIC ELECTRON

Consider the hydrogen atom with a potential $V = e^2/r$ and a wave function $\phi_0(\mathbf{R}) \exp(-iE_0 t) = \phi_0(x_\mu)$. Take the wave function to be normalized in the conventional manner. The effect of the self-energy of the electron is to shift the energy level by an amount

$$\Delta E = e^2 \int \tilde{\phi}_0(x_2,t_2)\gamma_\mu K_+{}^V(2,1)\gamma_\mu \delta_+(s_{1,2}{}^2)\,\phi_0(x_1,t_1)\,d^3x_1\,d^3x_2\,dt_2$$
$$-\Delta m \int \tilde{\phi}(x,t)\,\phi(x,t)\,d^3x \tag{30-6}$$

The first integral is written down from Fig. 30-1. The second is the free-particle effect as noted in previous lectures. The kernel $K_+{}^V$ is not well

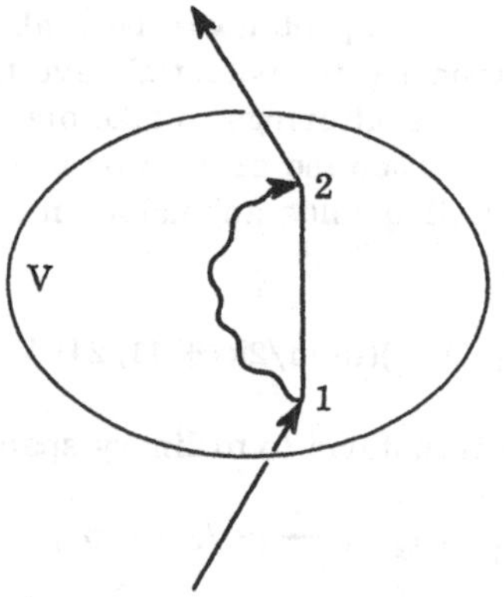

FIG. 30-1

enough determined to make exact calculation of this integral possible. An approximate calculation can be made with the form

$$K_+{}^V(2,1) = \sum_{+n} \exp[-iE_0(t_2 - t_1)]\,\tilde{\phi}_n(x_2)\phi_n(x_1) \qquad t_2 > t_1$$

similar sum over negative energies for $t_2 < t_1$

The photon propagation kernel can be expanded as

$$\delta_+(s_{1,2}^2) = 4\pi \int \exp[-ik(t_2 - t_1) + ik(x_2 - x_1)]\ d^3k/2k(2\pi)^{-3} \qquad t_2 > t_1$$

$$= 4\pi \int \exp[+ik(t_2 - t_1) + ik(x_2 - x_1)]\ d^3k/2k(2\pi)^{-3} \qquad t_2 < t_1$$

Using these expressions, Eq. (30-6) becomes

$$\Delta E = \sum_{+n} \int [\alpha_\mu \exp(-i\mathbf{K}\cdot\mathbf{R})]_{0n} (E_n + K - E_0)^{-1} [\alpha_\mu \exp(i\mathbf{K}\cdot\mathbf{R})]_{n0} \times d^3k/4\pi k - \sum_{-n} \int [\alpha_\mu \exp(-i\mathbf{K}\cdot\mathbf{R})]_{0n} (|E_n| + \omega + E_0)^{-1} \times [\alpha_\mu \exp(i\mathbf{K}\cdot\mathbf{R})]_{n0}\, d^3k/4\pi k - (\Delta m \text{ term}) \tag{30-7}$$

This form implies the use of ϕ^* instead of $\tilde{\phi}$ and $\alpha_4 = 1$, $\alpha_{1,2,3} = \boldsymbol{\alpha}$.

Another approach to the motion of an electron in a hydrogen atom is the following. Consider the electron as a free particle intermittently scattered by the Coulomb potential. The scatterings cause a phase shift in the wave function of the order of (Rydberg/$\hbar$). Thus the period between scatterings is of the order $T = \hbar/\text{Rydberg}$. Take the lower limit k_0 of the momentum of the "self-action" photons as very large compared to the Rydberg. Then it is very probable that an emitted photon will be reabsorbed before two interactions between the electron and the potential have taken place; it is very improbable for two or more scatterings to take place between emission and absorption (see Fig. 30-2). Then the correction to the potential is that computed in Eq. (30-5) for small q (plus anomalous moment correction). This is

$$(e^2/4\pi)(4q^2/3m^2)(\ln m/2k_0 + 11/24)\, \not{V}$$

in momentum space. To transform to ordinary space, use

$$q^2 \not{V} = (q_4^2 - Q^2)\not{V} \rightarrow (\partial^2/\partial t^2 - \nabla^2) V$$

Thus the correction is

$$-(e^2/3\pi m^2)(\log M/2k_0 + 11/24)\nabla^2 V \tag{30-7'}$$

This correction is of greatest importance for the s state, since with a Coulomb potential $\nabla^2 V = 4\pi Ze^2 \delta(\mathbf{R})$, and only in the s states is $\phi(\mathbf{R})$ different from 0 at $\mathbf{R} = 0$.

The choice of k_0 is determined by the inequalities $m \gg k_0 \gg$ Rydberg. A satisfactory value is $k_0 = 137$ Ryd. With such a k_0, the effect of photons of $k < k_0$ must be included. This will be done by separating the effect into the sum of three contributing effects. It will be seen that two of these effects

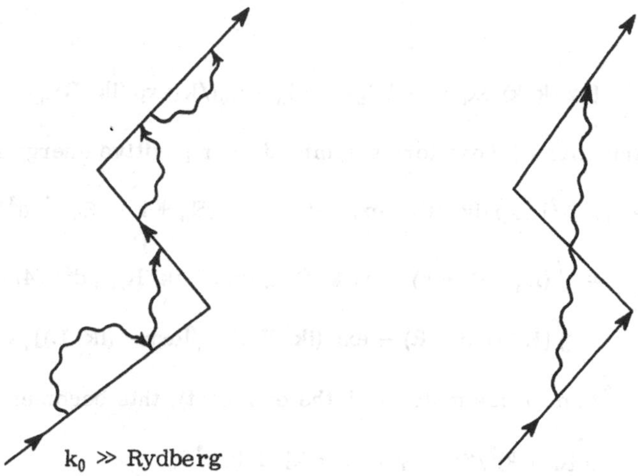

FIG. 30-2

are independent of the potential V and thus are canceled by similar terms in the Δm correction for a free particle. Thus for only one situation must the effect be computed. In all cases, since k is small, the nonrelativistic approximation to expression (30-7) may be used.

(1) The contribution of negative energy states: Neglecting k with respect to m gives

$$(|E_n| + k + E_0) \approx 2m$$

The matrix element for α_4 is very small, and only the elements for $\boldsymbol{\alpha}$ need be considered. Then the sum over negative states is

$$\sum_{-n} \int [(\boldsymbol{\alpha}_{0n}) \cdot (\boldsymbol{\alpha}_{n0})/2m]\, k^2\, dk/k$$

If this sum is continued for +n, a negligible term of order v^2/c^2 is added. Thus the sum is approximately

$$- \sum_{\text{all states}} \int [(\alpha_{0n}) \cdot (\alpha_{n0})/2m]\; k^2\, dk/k = (\alpha \cdot \alpha)_{00}\, k^2\, dk/2mk$$

$$= 3k_0{}^2/4m$$

This is independent of V, and thus is canceled by a similar quantity in the Δm term.

(2) Longitudinal positive energy states ($\alpha_\mu \rightarrow \boldsymbol{\alpha} \cdot \mathbf{k}/k$): As an exercise the reader may show

$$\boldsymbol{\alpha} \cdot \mathbf{k} \exp(i\mathbf{k} \cdot \mathbf{R}) = H \exp(i\mathbf{k} \cdot \mathbf{R}) - \exp(i\mathbf{k} \cdot \mathbf{R})H$$

Then

$$[(\boldsymbol{\alpha}\cdot\mathbf{k}/k)\exp(i\mathbf{k}\cdot\mathbf{R})]_{n0} = (E_n - E_0)/k[\exp(i\mathbf{k}\cdot\mathbf{R})]_{n0}$$

and the contribution of these terms summed over positive energy states gives

$$\int[1-(E_n-E_0)^2/k^2]\exp(i\mathbf{k}\cdot\mathbf{R})_{0n}\exp(-i\mathbf{k}\cdot\mathbf{R})_{n0}(E_n+k-E_0)^{-1}\,d^3k/4\pi k$$

$$=\int(E_n-E_0+k)\exp(i\mathbf{k}\cdot\mathbf{R})_{0n}\exp(-i\mathbf{k}\cdot\mathbf{R})_{n0}\,d^3k/4\pi k^3$$

$$=\int[H\exp(i\mathbf{k}\cdot\mathbf{R})-\exp(i\mathbf{k}\cdot\mathbf{R})H]_{0n}[\exp-(i\mathbf{k}\cdot\mathbf{R})]_{n0}\,d^3k/4\pi k^3$$

Writing $H = p^2/2m$ (V commutes with the exponent), this becomes

$$\int[(p+k)^2/2m - p^2/2m + k]\,d^3k/k^3$$

This term is independent of V, and thus is also canceled by the Δm correction.

(3) Transverse positive energy states: Since k_0 is large compared to the size of the atom, the dipole approximation can be used.† The general term in the sum of Eq. (30-7) becomes

$$\int(\alpha_{tr})_{0n}(\alpha_{tr})_{n0}(E_n+k-E_0)^{-1}\,d^3k/k \tag{30-8}$$

Writing

$$(E_n+k-E_0)^{-1} = 1/k - (E_n-E_0)/(E_n+k-E_0)k$$

the term in $1/k$ can be split off from the rest of the integral as a quantity independent of V and thus canceled by the Δm correction. Further, by averaging over directions,

$$(\alpha_{tr})_{0n}(\alpha_{tr})_{n0} = 2/3(\boldsymbol{\alpha})_{0n}\cdot(\boldsymbol{\alpha})_{n0} = (2/3m^2)(\mathbf{p})_{0n}\cdot(\mathbf{p})_{n0}$$

in the nonrelativistic approximation. Thus the integral of Eq. (30-8) is

$$(2/3m^2)(\mathbf{p})_{0n}\cdot(\mathbf{p})_{n0}(E_n-E_0)\log(k_0+E_n-E_0)/(E_n-E_0)$$

Using the relation

$$\mathbf{p}_{n0}(E_n-E_0) = (\mathbf{p}H - H\mathbf{p})_{n0} = (\nabla V)_{n0}$$

† Cf. H. Bethe, Phys. Rev., **72**, 339 (1947).

and the fact that $k_0 \gg E_n - E_0$, one part of the sum over transverse positive energy states is

$$\ln\ k_0 \sum_n \mathbf{p}_{0n} \cdot (\nabla V)_{n0} = 1/2\ \ln\ k_0 (\nabla^2 V)_{00}$$

This cancels with the $\ln\ k_0$ of Eq. (30-7′), leaving the final correction as

$$(2e^2/3\pi m^2) \sum_{+n} \mathbf{p}_{n0} \cdot \mathbf{p}_{0n} (E_n - E_0) \{ \log [M/2(E_n - E_0)] + (11/24) \}$$

$$+ \text{ anomalous moment correction}$$

This sum has been carried out numerically to be compared with the observed Lamb shift.

Thirty-first Lecture

CLOSED-LOOP PROCESSES, VACUUM POLARIZATION

Another process which is still of first order in e^2 has not been considered in the scattering by a potential. Instead of the potential scattering the particle directly, it can do so by first creating a pair which subsequently annihilates, creating a photon which does the scattering. Diagram I (Fig. 31-1) applies to this process; diagram II applies to a similar process, with the order in time changed slightly. The amplitude for these processes is

$$i4\pi e^2 \sum_{\substack{\text{spin states} \\ \text{of u}}} (\tilde{u}_2 \gamma_\mu u_1) \frac{1}{q^2} \int \left(\tilde{u} \frac{1}{\not{p} - m} \gamma_\mu \frac{1}{\not{p} + \not{q} - m} \not{a} u \right) (2\pi)^{-4}\, d^4p \qquad (31\text{-}1)$$

where u is the spinor part of the closed-loop wave function. The first parenthesis is the amplitude for the electron to be scattered by the photon; $1/q^2$ is the photon propagation factor; and the second parenthesis is the amplitude for the closed-loop process which produces the photon. The expression is integrated over p because the amplitude for a positron of <u>any</u> momenta is desired. In the sum over four spin states of u, two states take care of the processes of diagram I and two states take care of the processes of diagram II. No projection operators are required, so the method of spurs may be used directly to give

$$i4\pi e^2 (\tilde{u}_2 \gamma_\mu u_1) \frac{1}{q^2} \int \mathrm{Sp} \left[\frac{1}{\not{p} - m} \gamma_\mu \frac{1}{\not{p} + \not{q} - m} \not{a} \right] (2\pi)^{-4}\, d^4p \qquad (31\text{-}2)$$

a form which contains both I and II (so as usual it is not necessary to make separate diagrams for processes whose only difference is the order in time).

This integral also diverges, but a photon convergence factor, as used in the previous lectures, is of no value because now the integral is over p, the momentum of the positron in the intermediate step. The method which has been used to circumvent the divergence difficulty is to subtract from this integral, a similar integral with m replaced by M. M is taken to be much larger

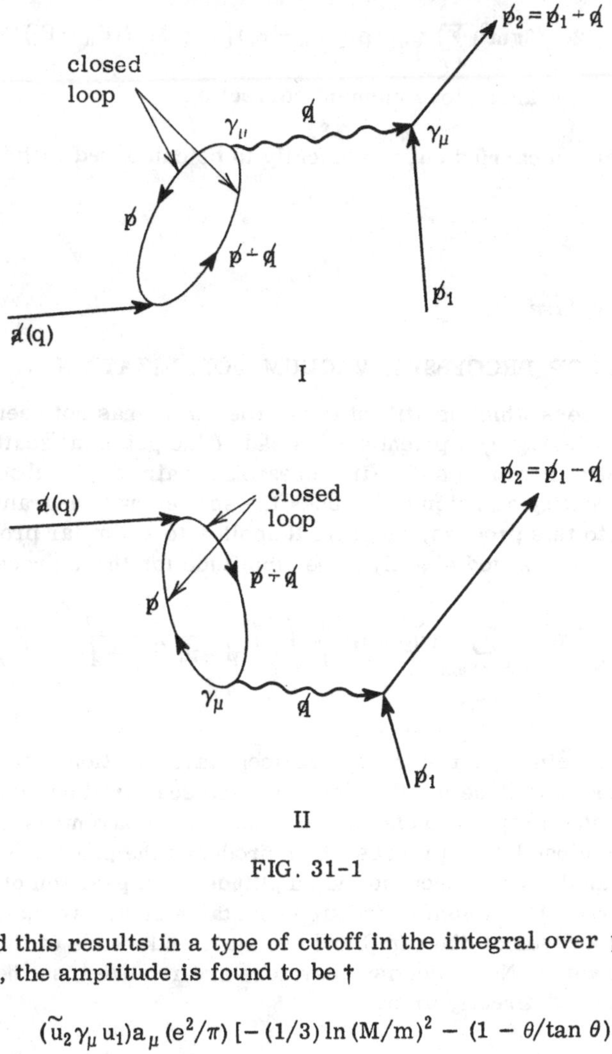

FIG. 31-1

than m, and this results in a type of cutoff in the integral over p. When this is done, the amplitude is found to be †

$$(\tilde{u}_2 \gamma_\mu u_1) a_\mu (e^2/\pi) [-(1/3) \ln (M/m)^2 - (1 - \theta/\tan\theta) \times (4m^2 + 2q^2)/3q^2 + 1/9] \tag{31-3}$$

†See R. P. Feynman, Phys. Rev., 76, 769 (1949); included in this volume.

where $q^2 = 4m^2 \sin^2 \theta$, which, for small q, becomes

$$(\tilde{u}_2 \gamma_\mu u_1) a_\mu \, (e^2/\pi)[-(1/3)\ln(M/m)^2 + 2q^2/15] \tag{31-4}$$

Notice that $(\tilde{u}_2 \gamma_\mu u_1) = (\tilde{u}_2 \not{a} u_1)$, so that, considering only the divergent part of the correction, the effective potential is

$$\not{a}\{1 + (e^2/\pi)[-(1/3)\ln(M/m)^2]\} \tag{31-5}$$

The 1 comes from the theory without radiative corrections, while the e^2 term is the correction due to processes of the type just described. Thus the correction can be interpreted as a small reduction in the effect of all potentials, and one can introduce an experimental charge e_{exp} and a theoretical charge e_{th} related by

$$e_{exp} = e_{th} + \Delta e \tag{31-6}$$

where $\Delta(e^2) = -(e^2/3\pi)\ln(M/m)^2$, in a manner analogous to the mass correction described in Lecture 28. This is referred to as ''charge renormalization.'' The other term,

$$(2/15)(e^2/\pi)q^2\not{a}$$

is more interesting, since it represents a perturbation $2e^2/15\pi(\nabla^2 V)$. This correction is responsible for 27 Mc in the Lamb shift and the $\{\ln[m/2(E_n - E_0)] + (11/24)\}$ term in (30-7′) is replaced by $\{\ln[m/2(E_n - E_0)] + (11/24) - (1/5)\}$. The 1/5 term is due to the ''polarization of the vacuum.''

SCATTERING OF LIGHT BY A POTENTIAL

One possible process for the scattering of light, and an indistinguishable alternative, is indicated by the diagrams in Fig. 31-2. The second diagram differs from the first only in the direction of the arrows of the electron lines. Reversing such a direction is equivalent to changing an electron to a positron. Thus the coupling with each potential would change sign. Since there are three such couplings, the amplitude for the second process is the negative of that for the first. Since the amplitudes add, the net amplitude is zero. In general, any closed-loop process of this type involving an odd number of couplings to a potential (including photon), has zero net amplitude.

Problem: Set up the integrals for each of the two diagrams in Fig. 31-2 and show that they are equal and opposite in sign.

However, the higher-order processes shown in Fig. 31-3 can take place. The amplitude for the process is

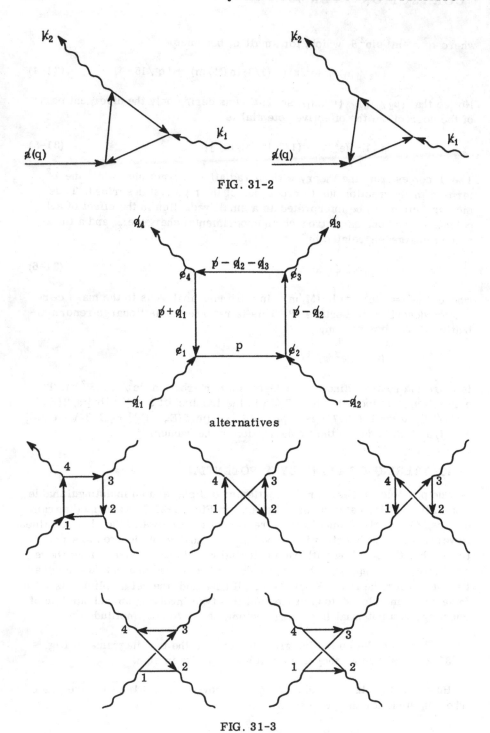

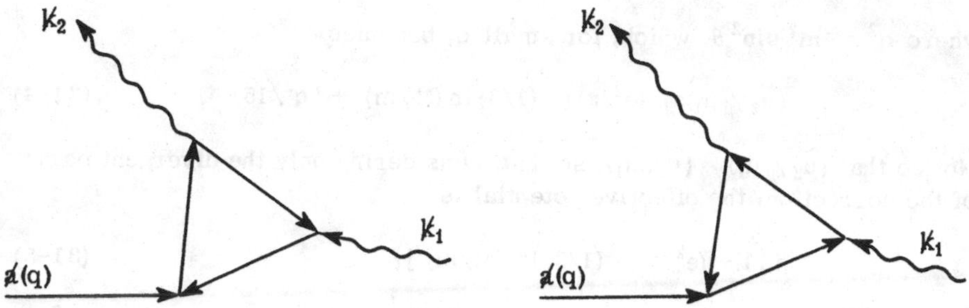

FIG. 31-2

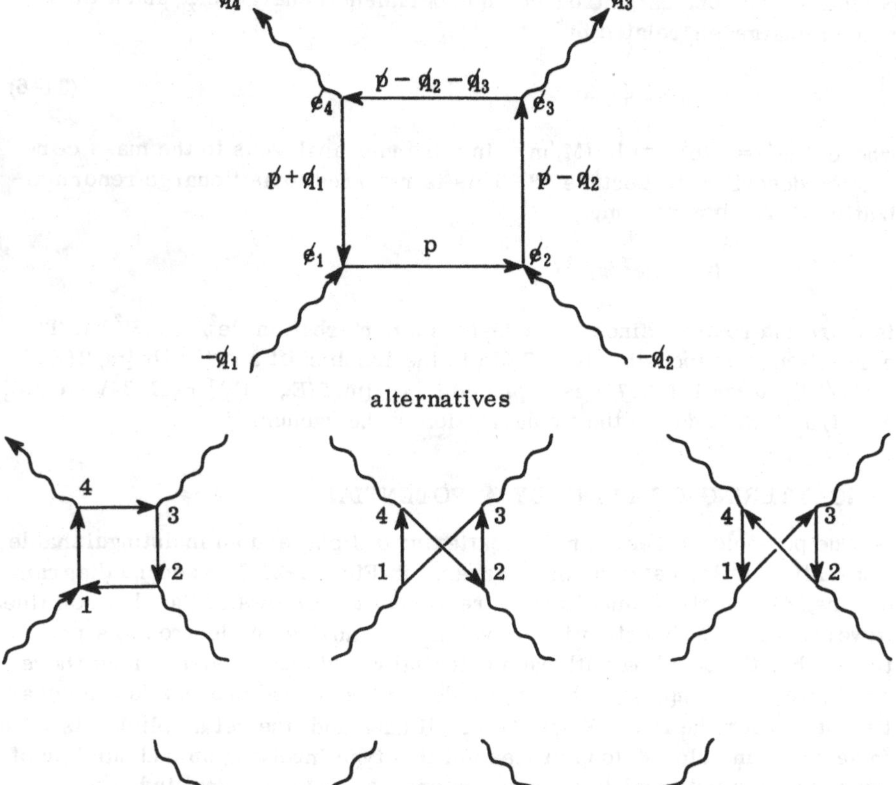

FIG. 31-3

$$-(4\pi e^2)^2 \int \mathrm{Sp}\,[\not{e}_1(\not{p}-m)^{-1}\,\not{e}_2(\not{p}_1-\not{q}_2-m)^{-1}\,\not{e}_3(\not{p}-\not{q}_2-\not{q}_3-m)^{-1}$$

$$\times\,\not{e}_4(\not{p}+\not{q}_1-m)^{-1}]\,(2\pi)^{-4}\,d^4k$$

plus five similar terms resulting from permuting the order of photons. This integral appears to diverge logarithmically. But when all six alternatives are taken into account, the sum leaves no divergent term. More complicated closed-loop processes are convergent.

Pauli Principle and the Dirac Equation

In Lecture 24 the probability of a vacuum remaining a vacuum under the influence of a potential was calculated. The potential can create and annihilate pairs (a closed-loop process) between times t_1 and t_2. The amplitude for the creation and annihilation of one pair is (to first nonvanishing order)

$$L \sim \int\int \mathrm{Sp}\,[K_+(1,2)\,\not{A}(2)\,K_+(2,1)\,\not{A}(1)]\;d\tau_1\,d\tau_2$$

The amplitude for the creation and annihilation for two pairs is a factor L for each, but, to avoid counting each twice when integrating over all $d\tau_1$ and $d\tau_2$, it is $L^2/2$. For three pairs the amplitude is $L^3/3!$. The total amplitude for a vacuum to remain a vacuum is, then,

$$c_v = 1 - L + L^2/2! - L^3/3! + \cdots = e^{-L} \qquad (31\text{-}7)$$

where the 1 comes from the amplitude to remain a vacuum with nothing happening. The use of minus signs for the amplitude for an odd number of pairs can be given the following justification in terms of the Pauli principle. Suppose the diagram for $t < t_1$ is as shown in Fig. 31-4. The completion of this process can occur in two ways, however (see Fig. 31-5). The second way can be thought of as obtained by the interchange of the two electrons, hence the amplitude of the second must be subtracted from that of the first,

FIG. 31-4

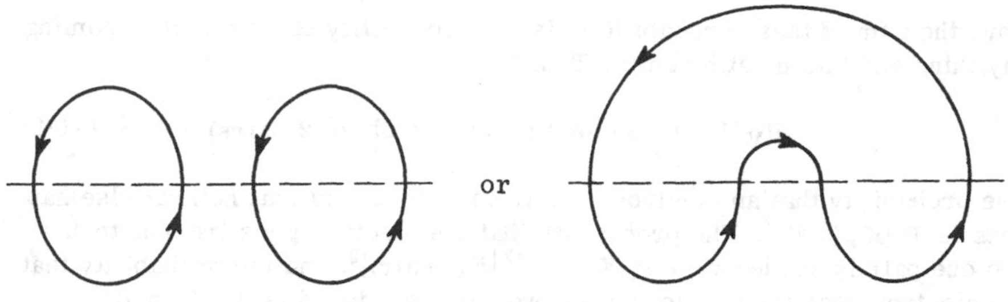

FIG. 31-5

according to the Pauli principle. But the second process is a one-loop process, whereas the first process is a two-loop process, so it can be concluded that amplitudes for an odd number of loops must be subtracted. The probability for a vacuum to remain a vacuum is

$$P_{\text{vac-vac}} = |c_v|^2 = \exp(-2 \text{ real part of } L)$$

The real part of L (R.P. of L) may be shown to be positive, so it is clear that terms of the series *must* alternate in sign in order that this probability be not greater than unity.

We have, therefore, two arguments as to why the expression must be e^{-L}. One involves the sign of the real part, a property just of K_+ and the Dirac equation. The second involves the Pauli principle. We see, therefore, that it could not be consistent to interpret the Dirac equation as we do unless the electrons obey Fermi-Dirac statistics. There is, therefore, some connection between the relativistic Dirac equation and the exclusion principle. Pauli has given a more elaborate proof of the necessity for the exclusion principle but this argument makes it plausible.

This question of the connection between the exclusion principle and the Dirac equation is so interesting that we shall try to give another argument that does not involve closed loops. We shall prove that it is inconsistent to assume that electrons are completely independent and wave functions for several electrons are simply products of individual wave functions (even though we neglect their interaction). For if we assume this, then

$$\left.\begin{array}{l}\text{Probability of vacuum}\\ \text{remaining a vacuum}\end{array}\right\} = P_V$$

$$\left.\begin{array}{l}\text{Probability of vacuum}\\ \text{to 1 pair}\end{array}\right\} = P_V \sum_{\text{all pairs}} |K_{1\,\text{pair}}|^2$$

$$\left.\begin{array}{l}\text{Probability of vacuum}\\ \text{to 2 pairs}\end{array}\right\} = P_V \sum_{\text{all pairs}} |K_{1\,\text{pair}}|^2\, |K_{1\,\text{pair}}|^2$$

⋮

Now, the sum of these probabilities is the probability of a vacuum becoming any thing and this must be unity. Thus

$$1 = P_V\,[1 + (\text{prob. of 1 pair}) + (\text{prob. of 2 pairs}) + \cdots] \quad (31\text{-}8)$$

The probability that an electron goes from a to b and that nothing else happens is $P_v|K_+(b,a)|^2$. The probability that the electron goes from a to b and one pair is produced is $P_v|K_+(b,a)|^2\,|K(1\text{ pair})|^2$, and the probability that the electron goes from a to b with two pairs produced is $P_v|K_+(b,a)|^2 \times |K(2\text{ pair})|^2$. Thus the probability for an electron to go from a to b with any number of pairs produced is

$$P_v|K_+(b,a)|^2\,[1 + |K(1\text{ pair})|^2 + |K(2\text{ pairs})|^2 + \cdots = |K_+(b,a)|^2 \quad (31\text{-}9)$$

[see Eq. (31-8)]. Now since the electron must go somewhere,

$$\int |K_+(b,a)|^2\,db = 1$$

However, it is a property of the Dirac kernel that

$$\int |K_+(b,a)|^2\,db > 1 \quad (31\text{-}10)$$

and an inconsistency results. The inconsistency can be eliminated by assuming that electrons obey Fermi-Dirac statistics and are not independent. Under these circumstances the original electron and the electron of the pair are not independent and

$$\left\{\begin{matrix}\text{Probability of electron from}\\ \text{a to b plus 1 pair produced}\end{matrix}\right\} < |K_+(b,a)|^2\,|K(1\text{ pair})|^2 \quad (31\text{-}11)$$

because we should not allow the case that the electron in the pair is in the same state as the electron at b.

For the kernel of the Klein-Gordon equation, it turns out that the sign of the inequality in Eq. (31-10) is reversed. Therefore, for a spin-zero particle neither Fermi-Dirac statistics nor independent particles are possible. If the wave functions are taken symmetric (charges reversed *add* amplitudes, Einstein-Bose statistics), the inequality Eq. (31-11) is also reversed. In symmetrical statistics the presence of a particle in a state (say 6) *enhances* the chance that another is created in the same state. So the Klein-Gordon equation requires Bose statistics.

It would be interesting to try to sharpen these arguments to show that the difference between $\int |K_+(b,a)|^2\,db$ and 1 is quantitatively exactly compensated for by the exclusion principle. Such a fundamental relation ought to have a clear and simple exposition.

10. SUMMARY OF NUMERICAL FACTORS FOR TRANSITION PROBABILITIES

The exact values of the numerical factors appearing in the rules of II for computing transition probabilities are not clearly stated there, so we give a brief summary here.[20]

The probability of transition per second from an initial state of energy E to a final state of the same total energy (assumed to be in a continuum) is given by

$(\hbar=c=1)$,

$$\text{Prob. trans/sec}=2\pi N^{-1}|\mathfrak{M}|^2\rho(E),$$

where $\rho(E)$ is the density of final states per unit energy range at energy E and $|\mathfrak{M}|^2$ is the square of the matrix element taken between the initial and final state of the transition matrix $\mathfrak{M}$ appropriate to the problem. N is a normalizing constant. For bound states conventionally normalized it is 1. For free particle states it is a product of a factor N_i for each particle in the initial and for each in the final energy state. N_i depends on the normalization of the wave functions of the particles (photons are considered as particles) which is used in computing the matrix element of $\mathfrak{M}$. The simplest rule (which does not destroy the apparent covariance of $\mathfrak{M}$), is[21] $N_i=2\epsilon_i$, where ϵ_i is the energy of the particle. This corresponds to choosing in momentum space, plane waves for photons of unit vector potential, $e^2=-1$. For electrons it corresponds to using $(\bar{u}u)=2m$ (so that, for example, if an electron is deviated from initial p_1 to final p_2, the sum over all initial and final spin states of $|\mathfrak{M}|^2$ is $\text{Sp}[(p_2+m)\mathfrak{M}(p_1+m)\bar{\mathfrak{M}}]$). Choice of normalization $(\bar{u}\gamma_t u)=1$ results in $N_i=1$ for electrons. The matrix $\mathfrak{M}$ is evaluated by making the diagrams and following the rules of II, but with the following definition of numerical factors. (We give them here for the special case that the initial, final, and intermediate states consist of free particles. The momentum space representation is then most convenient.)

First, write down the matrix directly without numerical factors. Thus, electron propagation factor is $(p-m)^{-1}$, virtual photon factor is k^{-2} with couplings $\gamma_\mu\cdots\gamma_\mu$. A real photon of polarization vector e_μ contributes factor e. A potential (times the electron charge, e) $A_\mu(x)$ contributes momentum q with amplitude $a(q)$, where $a_\mu(q)=\int A_\mu(1)\exp(iq\cdot x_1)d^4x_1$. (Note: On this point we deviate from the definition of a in I which is there $(2\pi)^{-2}$ times as large.) A spur is taken on the matrices of a closed loop. Because of the Pauli principle the sign is altered on contributions corresponding to an exchange of electron identity, and for each closed loop. One multiplies by $(2\pi)^{-4}d^4p=(2\pi)^{-4}dp_t dp_x dp_y dp_z$ and integrates over all values of any undetermined momentum variable p. (Note: On this point we again differ.[20])

The correct numerical value of $\mathfrak{M}$ is then obtained by multiplication by the following factors. (1) A factor $(4\pi)^{\frac{1}{2}}e$ for each coupling of an electron to a photon. Thus, a virtual photon, having two such couplings, contributes $4\pi e^2$. (In the units here, $e^2=1/137$ approximately and $(4\pi)^{\frac{1}{2}}e$ is just the charge on an electron in heaviside units.) (2) A further factor $-i$ for each virtual photon.

For meson theories the changes discussed in II, Sec. 10 are made in writing $\mathfrak{M}$, then further factors are (1) $(4\pi)^{\frac{1}{2}}g$ for each meson-nucleon coupling and (2) a factor $-i$ for each virtual spin one meson, but $+i$ for each virtual spin zero meson.

This suffices for transition probabilities, in which only the absolute square of $\mathfrak{M}$ is required. To get $\mathfrak{M}$ to be the actual phase shift per unit volume and time, additional factors of i for each virtual electron propagation, and $-i$ for each potential or photon interaction, are necessary. Then, for energy perturbation problems the energy shift is the expected value of $i\mathfrak{M}$ for the unperturbed state in question divided by the normalization constant N_i belonging to each particle comprising the unperturbed state.

The author has profited from discussions with M. Peshkin and L. Brown.

[20] In I and II the unfortunate convention was made that d^4k means $dk_4dk_1dk_2dk_3(2\pi)^{-2}$ for momentum space integrals. The confusing factor $(2\pi)^{-2}$ here serves no useful purpose, so the convention will be abandoned. In this section d^4k has its usual meaning, $dk_4dk_1dk_2dk_3$.

[21] In general, N_i is the particle density. It is $N_i=(\bar{u}\gamma_t u)$ for spin one-half fields and $i[(\phi^*\partial\phi/\partial t)-\phi\partial\phi^*/\partial t]$ for scalar fields. The latter is $2\epsilon_i$ if the field amplitude ϕ is taken as unity.

PHYSICAL REVIEW VOLUME 76, NUMBER 6 SEPTEMBER 15, 1949

The Theory of Positrons

R. P. Feynman
Department of Physics, Cornell University, Ithaca, New York

(Received April 8, 1949)

The problem of the behavior of positrons and electrons in given external potentials, neglecting their mutual interaction, is analyzed by replacing the theory of holes by a reinterpretation of the solutions of the Dirac equation. It is possible to write down a complete solution of the problem in terms of boundary conditions on the wave function, and this solution contains automatically all the possibilities of virtual (and real) pair formation and annihilation together with the ordinary scattering processes, including the correct relative signs of the various terms.

In this solution, the "negative energy states" appear in a form which may be pictured (as by Stückelberg) in space-time as waves traveling away from the external potential backwards in time. Experimentally, such a wave corresponds to a positron approaching the potential and annihilating the electron. A particle moving forward in time (electron) in a potential may be scattered forward in time (ordinary scattering) or backward (pair annihilation). When moving backward (positron) it may be scattered backward in time (positron scattering) or forward (pair production). For such a particle the amplitude for transition from an initial to a final state is analyzed to any order in the potential by considering it to undergo a sequence of such scatterings.

The amplitude for a process involving many such particles is the product of the transition amplitudes for each particle. The exclusion principle requires that antisymmetric combinations of amplitudes be chosen for those complete processes which differ only by exchange of particles. It seems that a consistent interpretation is only possible if the exclusion principle is adopted. The exclusion principle need not be taken into account in intermediate states. Vacuum problems do not arise for charges which do not interact with one another, but these are analyzed nevertheless in anticipation of application to quantum electrodynamics.

The results are also expressed in momentum-energy variables. Equivalence to the second quantization theory of holes is proved in an appendix.

1. INTRODUCTION

THIS is the first of a set of papers dealing with the solution of problems in quantum electrodynamics. The main principle is to deal directly with the solutions to the Hamiltonian differential equations rather than with these equations themselves. Here we treat simply the motion of electrons and positrons in given external potentials. In a second paper we consider the interactions of these particles, that is, quantum electrodynamics.

The problem of charges in a fixed potential is usually treated by the method of second quantization of the electron field, using the ideas of the theory of holes. Instead we show that by a suitable choice and interpretation of the solutions of Dirac's equation the problem may be equally well treated in a manner which is fundamentally no more complicated than Schrödinger's method of dealing with one or more particles. The various creation and annihilation operators in the conventional electron field view are required because the number of particles is not conserved, i.e., pairs may be created or destroyed. On the other hand charge is conserved which suggests that if we follow the charge, not the particle, the results can be simplified.

In the approximation of classical relativistic theory the creation of an electron pair (electron A, positron B) might be represented by the start of two world lines from the point of creation, 1. The world lines of the positron will then continue until it annihilates another electron, C, at a world point 2. Between the times t_1 and t_2 there are then three world lines, before and after only one. However, the world lines of C, B, and A together form one continuous line albeit the "positron part" B of this continuous line is directed backwards in time. Following the charge rather than the particles corresponds to considering this continuous world line as a whole rather than breaking it up into its pieces. It is as though a bombardier flying low over a road suddenly sees three roads and it is only when two of them come together and disappear again that he realizes that he has simply passed over a long switchback in a single road.

This over-all space-time point of view leads to considerable simplification in many problems. One can take into account at the same time processes which ordinarily would have to be considered separately. For example, when considering the scattering of an electron by a potential one automatically takes into account the effects of virtual pair productions. The same equation, Dirac's, which describes the deflection of the world line of an electron in a field, can also describe the deflection (and in just as simple a manner) when it is large enough to reverse the time-sense of the world line, and thereby correspond to pair annihilation. Quantum mechanically the direction of the world lines is replaced by the direction of propagation of waves.

This view is quite different from that of the Hamiltonian method which considers the future as developing continuously from out of the past. Here we imagine the entire space-time history laid out, and that we just become aware of increasing portions of it successively. In a scattering problem this over-all view of the complete scattering process is similar to the S-matrix viewpoint of Heisenberg. The temporal order of events during the scattering, which is analyzed in such detail by the Hamiltonian differential equation, is irrelevant. The relation of these viewpoints will be discussed much more fully in the introduction to the second paper, in which the more complicated interactions are analyzed.

The development stemmed from the idea that in nonrelativistic quantum mechanics the amplitude for a given process can be considered as the sum of an amplitude for each space-time path available.[1] In view of the fact that in classical physics positrons could be viewed as electrons proceeding along world lines toward the past (reference 7) the attempt was made to remove, in the relativistic case, the restriction that the paths must proceed always in one direction in time. It was discovered that the results could be even more easily understood from a more familiar physical viewpoint, that of scattered waves. This viewpoint is the one used in this paper. After the equations were worked out physically the proof of the equivalence to the second quantization theory was found.[2]

First we discuss the relation of the Hamiltonian differential equation to its solution, using for an example the Schrödinger equation. Next we deal in an analogous way with the Dirac equation and show how the solutions may be interpreted to apply to positrons. The interpretation seems not to be consistent unless the electrons obey the exclusion principle. (Charges obeying the Klein-Gordon equations can be described in an analogous manner, but here consistency apparently requires Bose statistics.)[3] A representation in momentum and energy variables which is useful for the calculation of matrix elements is described. A proof of the equivalence of the method to the theory of holes in second quantization is given in the Appendix.

2. GREEN'S FUNCTION TREATMENT OF SCHRÖDINGER'S EQUATION

We begin by a brief discussion of the relation of the non-relativistic wave equation to its solution. The ideas will then be extended to relativistic particles, satisfying Dirac's equation, and finally in the succeeding paper to interacting relativistic particles, that is, quantum electrodynamics.

The Schrödinger equation

$$i\partial\psi/\partial t = H\psi, \tag{1}$$

describes the change in the wave function ψ in an infinitesimal time Δt as due to the operation of an operator $\exp(-iH\Delta t)$. One can ask also, if $\psi(\mathbf{x}_1, t_1)$ is the wave function at $\mathbf{x}_1$ at time t_1, what is the wave function at time $t_2 > t_1$? It can always be written as

$$\psi(\mathbf{x}_2, t_2) = \int K(\mathbf{x}_2, t_2; \mathbf{x}_1, t_1)\psi(\mathbf{x}_1, t_1)d^3\mathbf{x}_1, \tag{2}$$

where K is a Green's function for the linear Eq. (1). (We have limited ourselves to a single particle of coordinate $\mathbf{x}$, but the equations are obviously of greater generality.) If H is a constant operator having eigenvalues E_n, eigenfunctions ϕ_n so that $\psi(\mathbf{x}, t_1)$ can be expanded as $\sum_n C_n\phi_n(\mathbf{x})$, then $\psi(\mathbf{x}, t_2) = \exp(-iE_n(t_2-t_1)) \times C_n\phi_n(\mathbf{x})$. Since $C_n = \int \phi_n^*(\mathbf{x}_1)\psi(\mathbf{x}_1, t_1)d^3\mathbf{x}_1$, one finds (where we write 1 for $\mathbf{x}_1$, t_1 and 2 for $\mathbf{x}_2$, t_2) in this case

$$K(2, 1) = \sum_n \phi_n(\mathbf{x}_2)\phi_n^*(\mathbf{x}_1) \exp(-iE_n(t_2-t_1)), \tag{3}$$

for $t_2 > t_1$. We shall find it convenient for $t_2 < t_1$ to define $K(2, 1) = 0$ (Eq. (2) is then not valid for $t_2 < t_1$). It is then readily shown that in general K can be defined by that solution of

$$(i\partial/\partial t_2 - H_2)K(2, 1) = i\delta(2, 1), \tag{4}$$

which is zero for $t_2 < t_1$, where $\delta(2, 1) = \delta(t_2-t_1)\delta(x_2-x_1) \times \delta(y_2-y_1)\delta(z_2-z_1)$ and the subscript 2 on H_2 means that the operator acts on the variables of 2 of $K(2, 1)$. When H is not constant, (2) and (4) are valid but K is less easy to evaluate than (3).[4]

We can call $K(2, 1)$ the total amplitude for arrival at $\mathbf{x}_2$, t_2 starting from $\mathbf{x}_1$, t_1. (It results from adding an amplitude, $\exp iS$, for each space time path between these points, where S is the action along the path.[1]) The transition amplitude for finding a particle in state $\chi(\mathbf{x}_2, t_2)$ at time t_2, if at t_1 it was in $\psi(\mathbf{x}_1, t_1)$, is

$$\int \chi^*(2)K(2, 1)\psi(1)d^3\mathbf{x}_1d^3\mathbf{x}_2. \tag{5}$$

A quantum mechanical system is described equally well by specifying the function K, or by specifying the Hamiltonian H from which it results. For some purposes the specification in terms of K is easier to use and visualize. We desire eventually to discuss quantum electrodynamics from this point of view.

To gain a greater familiarity with the K function and the point of view it suggests, we consider a simple perturbation problem. Imagine we have a particle in a weak potential $U(\mathbf{x}, t)$, a function of position and time. We wish to calculate $K(2, 1)$ if U differs from zero only for t between t_1 and t_2. We shall expand K in increasing powers of U:

$$K(2, 1) = K_0(2, 1) + K^{(1)}(2, 1) + K^{(2)}(2, 1) + \cdots. \tag{6}$$

To zero order in U, K is that for a free particle, $K_0(2, 1)$.[4] To study the first order correction $K^{(1)}(2, 1)$, first consider the case that U differs from zero only for the infinitesimal time interval Δt_3 between some time t_3 and $t_3 + \Delta t_3 (t_1 < t_3 < t_2)$. Then if $\psi(1)$ is the wave function at $\mathbf{x}_1$, t_1, the wave function at $\mathbf{x}_3$, t_3 is

$$\psi(3) = \int K_0(3, 1)\psi(1)d^3\mathbf{x}_1, \tag{7}$$

since from t_1 to t_3 the particle is free. For the short interval Δt_3 we solve (1) as

$$\begin{aligned}\psi(\mathbf{x}, t_3+\Delta t_3) &= \exp(-iH\Delta t_3)\psi(\mathbf{x}, t_3) \\ &= (1 - iH_0\Delta t_3 - iU\Delta t_3)\psi(\mathbf{x}, t_3),\end{aligned}$$

---

[1] R. P. Feynman, Rev. Mod. Phys. 20, 367 (1948).

[2] The equivalence of the entire procedure (including photon interactions) with the work of Schwinger and Tomonaga has been demonstrated by F. J. Dyson, Phys. Rev. 75, 486 (1949).

[3] These are special examples of the general relation of spin and statistics deduced by W. Pauli, Phys. Rev. 58, 716 (1940).

[4] For a non-relativistic free particle, where $\phi_n = \exp(i\mathbf{p}\cdot\mathbf{x})$, $E_n = \mathbf{p}^2/2m$, (3) gives, as is well known

$$\begin{aligned}K_0(2, 1) &= \int \exp[-(i\mathbf{p}\cdot\mathbf{x}_1 - i\mathbf{p}\cdot\mathbf{x}_2) - i\mathbf{p}^2(t_2-t_1)/2m]d^3\mathbf{p}(2\pi)^{-3} \\ &= (2\pi i m^{-1}(t_2-t_1))^{-\frac{3}{2}} \exp(\tfrac{1}{2}im(\mathbf{x}_2-\mathbf{x}_1)^2(t_2-t_1)^{-1})\end{aligned}$$

for $t_2 > t_1$, and $K_0 = 0$ for $t_2 < t_1$.

where we put $H=H_0+U$, H_0 being the Hamiltonian of a free particle. Thus $\psi(\mathbf{x}, t_3+\Delta t_3)$ differs from what it would be if the potential were zero (namely $(1-iH_0\Delta t_3)\psi(\mathbf{x}, t_3)$) by the extra piece

$$\Delta\psi = -iU(\mathbf{x}_3, t_3)\cdot\psi(\mathbf{x}_3, t_3)\Delta t_3, \quad (8)$$

which we shall call the amplitude scattered by the potential. The wave function at 2 is given by

$$\psi(\mathbf{x}_2, t_2) = \int K_0(\mathbf{x}_2, t_2; \mathbf{x}_3, t_3+\Delta t_3)\psi(\mathbf{x}_3, t_3+\Delta t_3)d^3\mathbf{x}_3,$$

since after $t_3+\Delta t_3$ the particle is again free. Therefore the change in the wave function at 2 brought about by the potential is (substitute (7) into (8) and (8) into the equation for $\psi(\mathbf{x}_2, t_2)$):

$$\Delta\psi(2) = -i\int K_0(2,3)U(3)K_0(3,1)\psi(1)d^3\mathbf{x}_1 d^3\mathbf{x}_3\Delta t_3.$$

In the case that the potential exists for an extended time, it may be looked upon as a sum of effects from each interval Δt_3 so that the total effect is obtained by integrating over t_3 as well as $\mathbf{x}_3$. From the definition (2) of K then, we find

$$K^{(1)}(2,1) = -i\int K_0(2,3)U(3)K_0(3,1)d\tau_3, \quad (9)$$

where the integral can now be extended over all space and time, $d\tau_3 = d^3\mathbf{x}_3 dt_3$. Automatically there will be no contribution if t_3 is outside the range t_1 to t_2 because of our definition, $K_0(2,1)=0$ for $t_2<t_1$.

We can understand the result (6), (9) this way. We can imagine that a particle travels as a free particle from point to point, but is scattered by the potential U. Thus the total amplitude for arrival at 2 from 1 can be considered as the sum of the amplitudes for various alternative routes. It may go directly from 1 to 2 (amplitude $K_0(2,1)$, giving the zero order term in (6)). Or (see Fig. 1(a)) it may go from 1 to 3 (amplitude $K_0(3,1)$), get scattered there by the potential (scattering amplitude $-iU(3)$ per unit volume and time) and then go from 3 to 2 (amplitude $K_0(2,3)$). This may occur for any point 3 so that summing over these alternatives gives (9).

Again, it may be scattered twice by the potential (Fig. 1(b)). It goes from 1 to 3 ($K_0(3,1)$), gets scattered there ($-iU(3)$) then proceeds to some other point, 4, in space time (amplitude $K_0(4,3)$) is scattered again ($-iU(4)$) and then proceeds to 2 ($K_0(2,4)$). Summing over all possible places and times for 3, 4 find that the second order contribution to the total amplitude $K^{(2)}(2,1)$ is

$$(-i)^2\int\int K_0(2,4)U(4)K_0(4,3) \times U(3)K_0(3,1)d\tau_3 d\tau_4. \quad (10)$$

This can be readily verified directly from (1) just as (9) was. One can in this way obviously write down any of the terms of the expansion (6).[5]

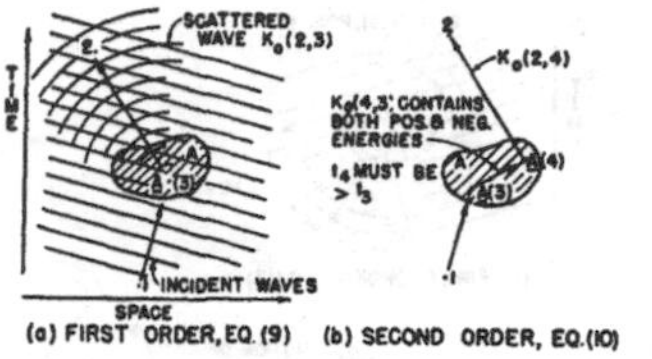

FIG. 1. The Schrödinger (and Dirac) equation can be visualized as describing the fact that plane waves are scattered successively by a potential. Figure 1 (a) illustrates the situation in first order. $K_0(2,3)$ is the amplitude for a free particle starting at point 3 to arrive at 2. The shaded region indicates the presence of the potential A which scatters at 3 with amplitude $-iA(3)$ per cm³sec. (Eq. (9)). In (b) is illustrated the second order process (Eq. (10)), the waves scattered at 3 are scattered again at 4. However, in Dirac one-electron theory $K_0(4,3)$ would represent electrons both of positive and of negative energies proceeding from 3 to 4. This is remedied by choosing a different scattering kernel $K_+(4,3)$, Fig. 2.

3. TREATMENT OF THE DIRAC EQUATION

We shall now extend the method of the last section to apply to the Dirac equation. All that would seem to be necessary in the previous equations is to consider H as the Dirac Hamiltonian, ψ as a symbol with four indices (for each particle). Then K_0 can still be defined by (3) or (4) and is now a 4–4 matrix which operating on the initial wave function, gives the final wave function. In (10), $U(3)$ can be generalized to $A_4(3)-\boldsymbol{\alpha}\cdot\mathbf{A}(3)$ where A_4, $\mathbf{A}$ are the scalar and vector potential (times e, the electron charge) and $\boldsymbol{\alpha}$ are Dirac matrices.

To discuss this we shall define a convenient relativistic notation. We represent four-vectors like $\mathbf{x}$, t by a symbol x_μ, where $\mu=1, 2, 3, 4$ and $x_4=t$ is real. Thus the vector and scalar potential (times e) $\mathbf{A}$, A_4 is A_μ. The four matrices $\beta\boldsymbol{\alpha}$, β can be considered as transforming as a four vector γ_μ (our γ_μ differs from Pauli's by a factor i for $\mu=1, 2, 3$). We use the summation convention $a_\mu b_\mu = a_4b_4 - a_1b_1 - a_2b_2 - a_3b_3 = a\cdot b$. In particular if a_μ is any four vector (but not a matrix) we write $\mathbf{a} = a_\mu\gamma_\mu$ so that $\mathbf{a}$ is a matrix associated with a vector ($\mathbf{a}$ will often be used in place of a_μ as a symbol for the vector). The γ_μ satisfy $\gamma_\mu\gamma_\nu+\gamma_\nu\gamma_\mu = 2\delta_{\mu\nu}$ where $\delta_{44}=+1$, $\delta_{11}=\delta_{22}=\delta_{33}=-1$, and the other $\delta_{\mu\nu}$ are zero. As a consequence of our summation convention $\delta_{\mu\nu}a_\nu = a_\mu$ and $\delta_{\mu\mu}=4$. Note that $\mathbf{ab}+\mathbf{ba} = 2a\cdot b$ and that $\mathbf{a}^2 = a_\mu a_\mu = a\cdot a$ is a pure number. The symbol $\partial/\partial x_\mu$ will mean $\partial/\partial t$ for $\mu=4$, and $-\partial/\partial x$, $-\partial/\partial y$, $-\partial/\partial z$ for $\mu=1, 2, 3$. Call $\nabla = \gamma_\mu\partial/\partial x_\mu = \beta\partial/\partial t + \beta\boldsymbol{\alpha}\cdot\boldsymbol{\nabla}$. We shall imagine

[5] We are simply solving by successive approximations an integral equation (deducible directly from (1) with $H=H_0+U$ and (4) with $H=H_0$),

$$\psi(2) = -i\int K_0(2,3)U(3)\psi(3)d\tau_3 + \int K_0(2,1)\psi(1)d^3\mathbf{x}_1,$$

where the first integral extends over all space and all times t_3 greater than the t_1 appearing in the second term, and $t_2>t_1$.

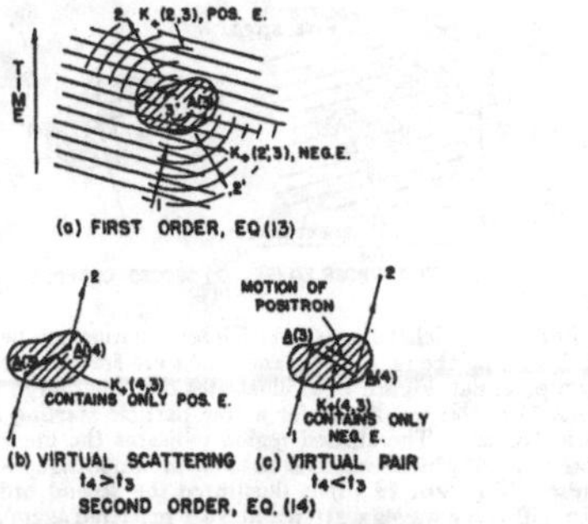

FIG. 2. The Dirac equation permits another solution $K_+(2, 1)$ if one considers that waves scattered by the potential can proceed backwards in time as in Fig. 2 (a). This is interpreted in the second order processes (b), (c), by noting that there is now the possibility (c) of virtual pair production at 4, the positron going to 3 to be annihilated. This can be pictured as similar to ordinary scattering (b) except that the electron is scattered backwards in time from 3 to 4. The waves scattered from 3 to 2′ in (a) represent the possibility of a positron arriving at 3 from 2′ and annihilating the electron from 1. This view is proved equivalent to hole theory: electrons traveling backwards in time are recognized as positrons.

hereafter, purely for relativistic convenience, that $\phi_n{}^*$ in (3) is replaced by its adjoint $\bar{\phi}_n=\phi_n{}^*\beta$.

Thus the Dirac equation for a particle, mass m, in an external field $A=A_\mu\gamma_\mu$ is

$$(i\nabla-m)\psi=A\psi, \qquad (11)$$

and Eq. (4) determining the propagation of a free particle becomes

$$(i\nabla_2-m)K_+(2,1)=i\delta(2,1), \qquad (12)$$

the index 2 on ∇_2 indicating differentiation with respect to the coordinates $x_{2\mu}$ which are represented as 2 in $K_+(2, 1)$ and $\delta(2, 1)$.

The function $K_+(2, 1)$ is defined in the absence of a field. If a potential A is acting a similar function, say $K_+{}^{(A)}(2, 1)$ can be defined. It differs from $K_+(2, 1)$ by a first order correction given by the analogue of (9) namely

$$K_+{}^{(1)}(2,1)=-i\int K_+(2,3)A(3)K_+(3,1)d\tau_3, \qquad (13)$$

representing the amplitude to go from 1 to 3 as a free particle, get scattered there by the potential (now the matrix $A(3)$ instead of $U(3)$) and continue to 2 as free. The second order correction, analogous to (10) is

$$K_+{}^{(2)}(2,1)=-\int\int K_+(2,4)A(4) \times K_+(4,3)A(3)K_+(3,1)d\tau_4 d\tau_3, \qquad (14)$$

and so on. In general $K_+{}^{(A)}$ satisfies

$$(i\nabla_2-A(2)-m)K_+{}^{(A)}(2,1)=i\delta(2,1), \qquad (15)$$

and the successive terms (13), (14) are the power series expansion of the integral equation

$$K_+{}^{(A)}(2,1)=K_+(2,1) -i\int K_+(2,3)A(3)K_+{}^{(A)}(3,1)d\tau_3, \qquad (16)$$

which it also satisfies.

We would now expect to choose, for the special solution of (12), $K_+=K_0$ where $K_0(2, 1)$ vanishes for $t_2<t_1$ and for $t_2>t_1$ is given by (3) where ϕ_n and E_n are the eigenfunctions and energy values of a particle satisfying Dirac's equation, and $\phi_n{}^*$ is replaced by $\bar{\phi}_n$.

The formulas arising from this choice, however, suffer from the drawback that they apply to the one electron theory of Dirac rather than to the hole theory of the positron. For example, consider as in Fig. 1(a) an electron after being scattered by a potential in a small region 3 of space time. The one electron theory says (as does (3) with $K_+=K_0$) that the scattered amplitude at another point 2 will proceed toward positive times with both positive and negative energies, that is with both positive and negative rates of change of phase. No wave is scattered to times previous to the time of scattering. These are just the properties of $K_0(2, 3)$.

On the other hand, according to the positron theory negative energy states are not available to the electron after the scattering. Therefore the choice $K_+=K_0$ is unsatisfactory. But there are other solutions of (12). We shall choose the solution defining $K_+(2, 1)$ so that *$K_+(2, 1)$ for $t_2>t_1$ is the sum of (3) over positive energy states only*. Now this new solution must satisfy (12) for all times in order that the representation be complete. It must therefore differ from the old solution K_0 by a solution of the homogeneous Dirac equation. It is clear from the definition that the difference K_0-K_+ is the sum of (3) over all negative energy states, as long as $t_2>t_1$. But this difference must be a solution of the homogeneous Dirac equation for all times and must therefore be represented by the same sum over negative energy states also for $t_2<t_1$. Since $K_0=0$ in this case, it follows that our new kernel, *$K_+(2, 1)$, for $t_2<t_1$ is the negative of the sum (3) over negative energy states*. That is,

$$\begin{aligned} K_+(2,1)&=\textstyle\sum_{POS\,E_n}\phi_n(2)\bar{\phi}_n(1) \times\exp(-iE_n(t_2-t_1)) \quad \text{for } t_2>t_1 \\ &=-\textstyle\sum_{NEG\,E_n}\phi_n(2)\bar{\phi}_n(1) \times\exp(-iE_n(t_2-t_1)) \quad \text{for } t_2<t_1. \end{aligned} \qquad (17)$$

With this choice of K_+ our equations such as (13) and (14) will now give results equivalent to those of the positron hole theory.

That (14), for example, is the correct second order expression for finding at 2 an electron originally at 1 according to the positron theory may be seen as follows (Fig. 2). Assume as a special example that $t_2>t_1$ and that the potential vanishes except in interval t_2-t_1 so that t_4 and t_3 both lie between t_1 and t_2.

First suppose $t_4>t_3$ (Fig. 2(b)). Then (since $t_3>t_1$)

the electron assumed originally in a positive energy state propagates in that state (by $K_+(3,1)$) to position 3 where it gets scattered ($A(3)$). It then proceeds to 4, which it must do as a positive energy electron. This is correctly described by (14) for $K_+(4,3)$ contains only positive energy components in its expansion, as $t_4 > t_3$. After being scattered at 4 it then proceeds on to 2, again necessarily in a positive energy state, as $t_2 > t_4$.

In positron theory there is an additional contribution due to the possibility of virtual pair production (Fig. 2(c)). A pair could be created by the potential $A(4)$ at 4, the electron of which is that found later at 2. The positron (or rather, the hole) proceeds to 3 where it annihilates the electron which has arrived there from 1.

This alternative is already included in (14) as contributions for which $t_4 < t_3$, and its study will lead us to an interpretation of $K_+(4,3)$ for $t_4 < t_3$. The factor $K_+(2,4)$ describes the electron (after the pair production at 4) proceeding from 4 to 2. Likewise $K_+(3,1)$ represents the electron proceeding from 1 to 3. $K_+(4,3)$ must therefore represent the propagation of the positron or hole from 4 to 3. That it does so is clear. The fact that in hole theory the hole proceeds in the manner of and electron of negative energy is reflected in the fact that $K_+(4,3)$ for $t_4 < t_3$ is (minus) the sum of only negative energy components. In hole theory the real energy of these intermediate states is, of course, positive. This is true here too, since in the phases $\exp(-iE_n(t_4-t_3))$ defining $K_+(4,3)$ in (17), E_n is negative but so is t_4-t_3. That is, the contributions vary with t_3 as $\exp(-i|E_n|(t_3-t_4))$ as they would if the energy of the intermediate state were $|E_n|$. The fact that the entire sum is taken as negative in computing $K_+(4,3)$ is reflected in the fact that in hole theory the amplitude has its sign reversed in accordance with the Pauli principle and the fact that the electron arriving at 2 has been exchanged with one in the sea.[6] To this, and to higher orders, all processes involving virtual pairs are correctly described in this way.

The expressions such as (14) can still be described as a passage of the electron from 1 to 3 ($K_+(3,1)$), scattering at 3 by $A(3)$, proceeding to 4 ($K_+(4,3)$), scattering again, $A(4)$, arriving finally at 2. The scatterings may, however, be toward both future and past times, an electron propagating backwards in time being recognized as a positron.

This therefore suggests that negative energy components created by scattering in a potential be considered as waves propagating from the scattering point toward the past, and that such waves represent the propagation of a positron annihilating the electron in the potential.[7]

[6] It has often been noted that the one-electron theory apparently gives the same matrix elements for this process as does hole theory. The problem is one of interpretation, especially in a way that will also give correct results for other processes, e.g., self-energy.

[7] The idea that positrons can be represented as electrons with proper time reversed relative to true time has been discussed by the author and others, particularly by Stückelberg. E. C. C. Stückelberg, Helv. Phys. Acta **15**, 23 (1942); R. P. Feynman, Phys. Rev. **74**, 939 (1948). The fact that classically the action (proper time) increases continuously as one follows a trajectory is reflected in quantum mechanics in the fact that the phase, which is $|E_n||t_2-t_1|$, always increases as the particle proceeds from one scattering point to the next.

With this interpretation real pair production is also described correctly (see Fig. 3). For example in (13) if $t_1 < t_3 < t_2$ the equation gives the amplitude that if at time t_1 one electron is present at 1, then at time t_2 just one electron will be present (having been scattered at 3) and it will be at 2. On the other hand if t_2 is less than t_3, for example, if $t_2 = t_1 < t_3$, the same expression gives the amplitude that a pair, electron at 1, positron at 2 will annihilate at 3, and subsequently no particles will be present. Likewise if t_2 and t_1 exceed t_3 we have (minus) the amplitude for finding a single pair, electron at 2, positron at 1 created by $A(3)$ from a vacuum. If $t_1 > t_3 > t_2$, (13) describes the scattering of a positron. All these amplitudes are relative to the amplitude that a vacuum will remain a vacuum, which is taken as unity. (This will be discussed more fully later.)

The analogue of (2) can be easily worked out.[8] It is,

$$\psi(2) = \int K_+(2,1)N(1)\psi(1)d^3V_1, \tag{18}$$

where d^3V_1 is the volume element of the closed 3-dimensional surface of a region of space time containing

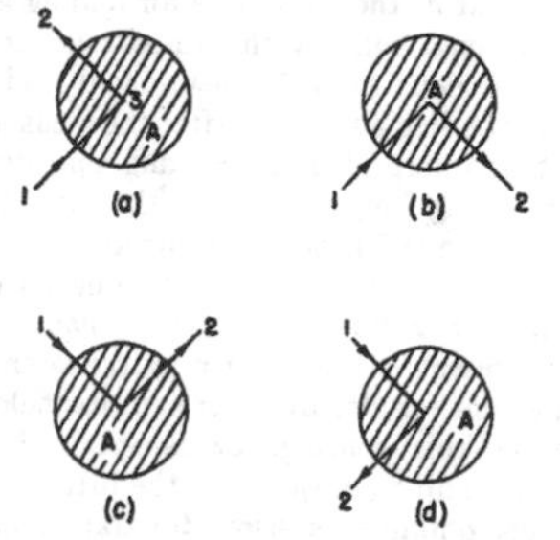

FIG. 3. Several different processes can be described by the same formula depending on the time relations of the variables t_2, t_1. Thus $P_v|K_+^{(A)}(2,1)|^2$ is the probability that: (a) An electron at 1 will be scattered at 2 (and no other pairs form in vacuum). (b) Electron at 1 and positron at 2 annihilate leaving nothing. (c) A single pair at 1 and 2 is created from vacuum. (d) A positron at 2 is scattered to 1. ($K_+^{(A)}(2,1)$ is the sum of the effects of scattering in the potential to all orders. P_v is a normalizing constant.)

[8] By multiplying (12) on the right by $(-i\nabla_1 - m)$ and noting that $\nabla_1\delta(2,1) = -\nabla_2\delta(2,1)$ show that $K_+(2,1)$ also satisfies $K_+(2,1)(-i\nabla_1 - m) = i\delta(2,1)$, where the ∇_1 operates on variable 1 in $K_+(2,1)$ but is written after that function to keep the correct order of the γ matrices. Multiply this equation by $\psi(1)$ and Eq. (11) (with $A=0$, calling the variables 1) by $K_+(2,1)$, subtract and integrate over a region of space-time. The integral on the left-hand side can be transformed to an integral over the surface of the region. The right-hand side is $\psi(2)$ if the point 2 lies within the region, and is zero otherwise. (What happens when the 3-surface contains a light line and hence has no unique normal need not concern us as these points can be made to occur so far away from 2 that their contribution vanishes.)

point 2, and $N(1)$ is $N_\mu(1)\gamma_\mu$ where $N_\mu(1)$ is the *inward* drawn unit normal to the surface at the point 1. That is, the wave function $\psi(2)$ (in this case for a free particle) is determined at any point inside a four-dimensional region if its values on the surface of that region are specified.

To interpret this, consider the case that the 3-surface consists essentially of all space at some time say $t=0$ previous to t_2, and of all space at the time $T>t_2$. The cylinder connecting these to complete the closure of the surface may be very distant from $\mathbf{x}_2$ so that it gives no appreciable contribution (as $K_+(2,1)$ decreases exponentially in space-like directions). Hence, if $\gamma_4=\beta$, since the inward drawn normals N will be β and $-\beta$,

$$\psi(2)=\int K_+(2,1)\beta\psi(1)d^3\mathbf{x}_1 - \int K_+(2,1')\beta\psi(1')d^3\mathbf{x}_{1'}, \quad (19)$$

where $t_1=0$, $t_{1'}=T$. Only positive energy (electron) components in $\psi(1)$ contribute to the first integral and only negative energy (positron) components of $\psi(1')$ to the second. That is, the amplitude for finding a charge at 2 is determined both by the amplitude for finding an electron previous to the measurement and by the amplitude for finding a positron after the measurement. This might be interpreted as meaning that even in a problem involving but one charge the amplitude for finding the charge at 2 is not determined when the only thing known in the amplitude for finding an electron (or a positron) at an earlier time. There may have been no electron present initially but a pair was created in the measurement (or also by other external fields). The amplitude for this contingency is specified by the amplitude for finding a positron in the future.

We can also obtain expressions for transition amplitudes, like (5). For example if at $t=0$ we have an electron present in a state with (positive energy) wave function $f(\mathbf{x})$, what is the amplitude for finding it at $t=T$ with the (positive energy) wave function $g(\mathbf{x})$? The amplitude for finding the electron anywhere after $t=0$ is given by (19) with $\psi(1)$ replaced by $f(\mathbf{x})$, the second integral vanishing. Hence, the transition element to find it in state $g(\mathbf{x})$ is, in analogy to (5), just $(t_2=T,\ t_1=0)$

$$\int \bar{g}(\mathbf{x}_2)\beta K_+(2,1)\beta f(\mathbf{x}_1)d^3\mathbf{x}_1d^3\mathbf{x}_2, \quad (20)$$

since $g^*=\bar{g}\beta$.

If a potential acts somewhere in the interval between 0 and T, K_+ is replaced by $K_+^{(A)}$. Thus the first order effect on the transition amplitude is, from (13),

$$-i\int \bar{g}(\mathbf{x}_2)\beta K_+(2,3)A(3)K_+(3,1)\beta f(\mathbf{x}_1)d^3\mathbf{x}_1d^3\mathbf{x}_2. \quad (21)$$

Expressions such as this can be simplified and the 3-surface integrals, which are inconvenient for relativistic calculations, can be removed as follows. Instead of defining a state by the wave function $f(\mathbf{x})$, which it has at a given time $t_1=0$, we define the state by the function $f(1)$ of four variables $\mathbf{x}_1$, t_1 which is a solution of the free particle equation for all t_1 and is $f(\mathbf{x}_1)$ for $t_1=0$. The final state is likewise defined by a function $g(2)$ over-all space-time. Then our surface integrals can be performed since $\int K_+(3,1)\beta f(\mathbf{x}_1)d^3\mathbf{x}_1=f(3)$ and $\int \bar{g}(\mathbf{x}_2)\beta d^3\mathbf{x}_2K_+(2,3)=\bar{g}(3)$. There results

$$-i\int \bar{g}(3)A(3)f(3)d\tau_3, \quad (22)$$

the integral now being over-all space-time. The transition amplitude to second order (from (14)) is

$$-\int\int \bar{g}(2)A(2)K_+(2,1)A(1)f(1)d\tau_1d\tau_2, \quad (23)$$

for the particle arriving at 1 with amplitude $f(1)$ is scattered $(A(1))$, progresses to 2, $(K_+(2,1))$, and is scattered again $(A(2))$, and we then ask for the amplitude that it is in state $g(2)$. If $g(2)$ is a negative energy state we are solving a problem of annihilation of electron in $f(1)$, positron in $g(2)$, etc.

We have been emphasizing scattering problems, but obviously the motion in a fixed potential V, say in a hydrogen atom, can also be dealt with. If it is first viewed as a scattering problem we can ask for the amplitude, $\phi_k(1)$, that an electron with original free wave function was scattered k times in the potential V either forward or backward in time to arrive at 1. Then the amplitude after one more scattering is

$$\phi_{k+1}(2)=-i\int K_+(2,1)V(1)\phi_k(1)d\tau_1. \quad (24)$$

An equation for the total amplitude

$$\psi(1)=\sum_{k=0}^{\infty}\phi_k(1)$$

for arriving at 1 either directly or after any number of scatterings is obtained by summing (24) over all k from 0 to ∞;

$$\psi(2)=\phi_0(2)-i\int K_+(2,1)V(1)\psi(1)d\tau_1. \quad (25)$$

Viewed as a steady state problem we may wish, for example, to find that initial condition ϕ_0 (or better just the ψ) which leads to a periodic motion of ψ. This is most practically done, of course, by solving the Dirac equation,

$$(i\nabla-m)\psi(1)=V(1)\psi(1), \quad (26)$$

deduced from (25) by operating on both sides by $i\nabla_2-m$, thereby eliminating the ϕ_0, and using (12). This illustrates the relation between the points of view.

For many problems the total potential $A+V$ may be split conveniently into a fixed one, V, and another, A, considered as a perturbation. If $K_+^{(V)}$ is defined as in

(16) with V for A, expressions such as (23) are valid and useful with K_+ replaced by $K_+^{(V)}$ and the functions $f(1)$, $g(2)$ replaced by solutions for all space and time of the Dirac Eq. (26) in the potential V (rather than free particle wave functions).

4. PROBLEMS INVOLVING SEVERAL CHARGES

We wish next to consider the case that there are two (or more) distinct charges (in addition to pairs they may produce in virtual states). In a succeeding paper we discuss the interaction between such charges. Here we assume that they do not interact. In this case each particle behaves independently of the other. We can expect that if we have two particles a and b, the amplitude that particle a goes from $\mathbf{x}_1$ at t_1, to $\mathbf{x}_3$ at t_3 while b goes from $\mathbf{x}_2$ at t_2 to $\mathbf{x}_4$ at t_4 is the product

$$K(3,4;1,2)=K_{+a}(3,1)K_{+b}(4,2).$$

The symbols a, b simply indicate that the matrices appearing in the K_+ apply to the Dirac four component spinors corresponding to particle a or b respectively (the wave function now having 16 indices). In a potential K_{+a} and K_{+b} become $K_{+a}^{(A)}$ and $K_{+b}^{(A)}$ where $K_{+a}^{(A)}$ is defined and calculated as for a single particle. They commute. Hereafter the a, b can be omitted; the space time variable appearing in the kernels suffice to define on what they operate.

The particles are identical however and satisfy the exclusion principle. The principle requires only that one calculate $K(3,4;1,2)-K(4,3;1,2)$ to get the net amplitude for arrival of charges at 3, 4. (It is normalized assuming that when an integral is performed over points 3 and 4, for example, since the electrons represented are identical, one divides by 2.) This expression is correct for positrons also (Fig. 4). For example the amplitude that an electron and a positron found initially at $\mathbf{x}_1$ and $\mathbf{x}_4$ (say $t_1=t_4$) are later found at $\mathbf{x}_3$ and $\mathbf{x}_2$ (with $t_2=t_3>t_1$) is given by the same expression

$$K_+^{(A)}(3,1)K_+^{(A)}(4,2)-K_+^{(A)}(4,1)K_+^{(A)}(3,2). \quad (27)$$

The first term represents the amplitude that the electron proceeds from 1 to 3 and the positron from 4 to 2 (Fig. 4(c)), while the second term represents the interfering amplitude that the pair at 1, 4 annihilate and what is found at 3, 2 is a pair newly created in the potential. The generalization to several particles is clear. There is an additional factor $K_+^{(A)}$ for each particle, and antisymmetric combinations are always taken.

No account need be taken of the exclusion principle in intermediate states. As an example consider again expression (14) for $t_2>t_1$ and suppose $t_4<t_3$ so that the situation represented (Fig. 2(c)) is that a pair is made at 4 with the electron proceeding to 2, and the positron to 3 where it annihilates the electron arriving from 1. It may be objected that if it happens that the electron created at 4 is in the same state as the one coming from 1, then the process cannot occur because of the exclusion principle and we should not have included it in our term (14). We shall see, however, that considering the exclusion principle also requires another change which reinstates the quantity.

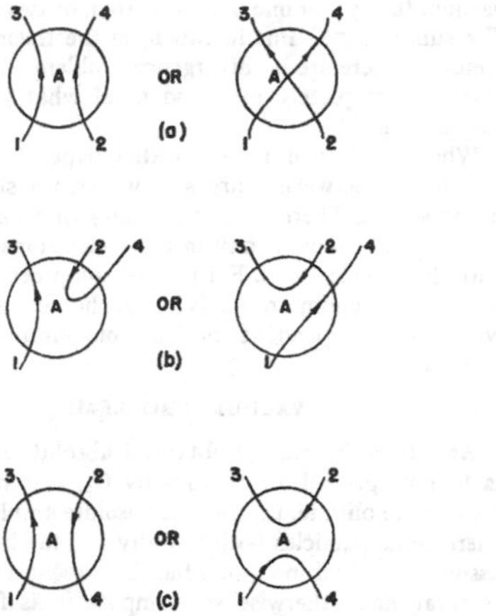

FIG. 4. Some problems involving two distinct charges (in addition to virtual pairs they may produce): $P_v|K_+^{(A)}(3,1)K_+^{(A)}(4,2)-K_+^{(A)}(4,1)K_+^{(A)}(3,2)|^2$ is the probability that: (a) Electrons at 1 and 2 are scattered to 3, 4 (and no pairs are formed). (b) Starting with an electron at 1 a single pair is formed, positron at 2, electrons at 3, 4. (c) A pair at 1, 4 is found at 3, 2, etc. The exclusion principle requires that the amplitudes for processes involving exchange of two electrons be subtracted.

For we are computing amplitudes relative to the amplitude that a vacuum at t_1 will still be a vacuum at t_2. We are interested in the alteration in this amplitude due to the presence of an electron at 1. Now one process that can be visualized as occurring in the vacuum is the creation of a pair at 4 followed by a re-annihilation of the *same* pair at 3 (a process which we shall call a closed loop path). But if a real electron is present in a certain state 1, those pairs for which the electron was created in state 1 in the vacuum must now be excluded. We must therefore subtract from our relative amplitude the term corresponding to this process. But this just reinstates the quantity which it was argued should not have been included in (14), the necessary minus sign coming automatically from the definition of K_+. It is obviously simpler to disregard the exclusion principle completely in the intermediate states.

All the amplitudes are relative and their squares give the relative probabilities of the various phenomena. Absolute probabilities result if one multiplies each of the probabilities by P_v, the true probability that if one has no particles present initially there will be none finally. This quantity P_v can be calculated by normalizing the relative probabilities such that the sum of the probabilities of all mutually exclusive alternatives is unity. (For example if one starts with a vacuum one can calculate the relative probability that there remains a vacuum (unity), or one pair is created, or two pairs, etc. The sum is P_v^{-1}.) Put in this form the theory is complete and there are no divergence problems. Real processes are completely independent of what goes on in the vacuum.

When we come, in the succeeding paper, to deal with interactions between charges, however, the situation is not so simple. There is the possibility that virtual electrons in the vacuum may interact electromagnetically with the real electrons. For that reason processes occurring in the vacuum are analyzed in the next section, in which an independent method of obtaining P_v is discussed.

5. VACUUM PROBLEMS

An alternative way of obtaining absolute amplitudes is to multiply all amplitudes by C_v, the vacuum to vacuum amplitude, that is, the absolute amplitude that there be no particles both initially and finally. We can assume $C_v=1$ if no potential is present during the interval, and otherwise we compute it as follows. It differs from unity because, for example, a pair could be created which eventually annihilates itself again. Such a path would appear as a closed loop on a space-time diagram. The sum of the amplitudes resulting from all such single closed loops we call L. To a first approximation L is

$$L^{(1)}=-\frac{1}{2}\int\int Sp[K_+(2,1)A(1)\times K_+(1,2)A(2)]d\tau_1 d\tau_2. \quad (28)$$

For a pair could be created say at 1, the electron and positron could both go on to 2 and there annihilate. The spur, Sp, is taken since one has to sum over all possible spins for the pair. The factor $\frac{1}{2}$ arises from the fact that the same loop could be considered as starting at either potential, and the minus sign results since the interactors are each $-iA$. The next order term would be[9]

$$L^{(2)}=+(i/3)\int\int\int Sp[K_+(2,1)A(1)\times K_+(1,3)A(3)K_+(3,2)A(2)]d\tau_1 d\tau_2 d\tau_3,$$

etc. The sum of all such terms gives L.[10]

In addition to these single loops we have the possibility that two independent pairs may be created and each pair may annihilate itself again. That is, there may be formed in the vacuum two closed loops, and the contribution in amplitude from this alternative is just the product of the contribution from each of the loops considered singly. The total contribution from all such pairs of loops (it is still consistent to disregard the exclusion principle for these virtual states) is $L^2/2$ for in L^2 we count every pair of loops twice. The total vacuum-vacuum amplitude is then

$$C_v=1-L+L^2/2-L^3/6+\cdots=\exp(-L), \quad (30)$$

the successive terms representing the amplitude from zero, one, two, etc., loops. The fact that the contribution to C_v of single loops is $-L$ is a consequence of the Pauli principle. For example, consider a situation in which two pairs of particles are created. Then these pairs later destroy themselves so that we have two loops. The electrons could, at a given time, be interchanged forming a kind of figure eight which is a single loop. The fact that the interchange must change the sign of the contribution requires that the terms in C_v appear with alternate signs. (The exclusion principle is also responsible in a similar way for the fact that the amplitude for a pair creation is $-K_+$ rather than $+K_+$.) Symmetrical statistics would lead to

$$C_v=1+L+L^2/2=\exp(+L).$$

The quantity L has an infinite imaginary part (from $L^{(1)}$, higher orders are finite). We will discuss this in connection with vacuum polarization in the succeeding paper. This has no effect on the normalization constant for the probability that a vacuum remain vacuum is given by

$$P_v=|C_v|^2=\exp(-2\cdot\text{real part of }L),$$

from (30). This value agrees with the one calculated directly by renormalizing probabilities. The real part of L appears to be positive as a consequence of the Dirac equation and properties of K_+ so that P_v is less than one. Bose statistics gives $C_v=\exp(+L)$ and consequently a value of P_v greater than unity which appears meaningless if the quantities are interpreted as we have done here. Our choice of K_+ apparently requires the exclusion principle.

Charges obeying the Klein-Gordon equation can be equally well treated by the methods which are discussed here for the Dirac electrons. How this is done is discussed in more detail in the succeeding paper. The real part of L comes out negative for this equation so that in this case Bose statistics appear to be required for consistency.[3]

---

[9] This term actually vanishes as can be seen as follows. In any spur the sign of all γ matrices may be reversed. Reversing the sign of γ in $K_+(2,1)$ changes it to the transpose of $K_+(1,2)$ so that the order of all factors and variables is reversed. Since the integral is taken over all τ_1, τ_2, and τ_3 this has no effect and we are left with $(-1)^3$ from changing the sign of A. Thus the spur equals its negative. Loops with an odd number of potential interactors give zero. Physically this is because for each loop the electron can go around one way or in the opposite direction and we must add these amplitudes. But reversing the motion of an electron makes it behave like a positive charge thus changing the sign of each potential interaction, so that the sum is zero if the number of interactions is odd. This theorem is due to W. H. Furry, Phys. Rev. 51, 125 (1937).

[10] A closed expression for L in terms of $K_+^{(A)}$ is hard to obtain because of the factor $(1/n)$ in the nth term. However, the perturbation in L, ΔL due to a small change in potential ΔA, is easy to express. The $(1/n)$ is canceled by the fact that ΔA can appear in any of the n potentials. The result after summing over n by (13), (14) and using (16) is

$$\Delta L=-i\int Sp[(K_+^{(A)}(1,1)-K_+(1,1))\Delta A(1)]d\tau_1. \quad (29)$$

The term $K_+(1,1)$ actually integrates to zero.

6. ENERGY-MOMENTUM REPRESENTATION

The practical evaluation of the matrix elements in some problems is often simplified by working with momentum and energy variables rather than space and time. This is because the function $K_+(2, 1)$ is fairly complicated but we shall find that its Fourier transform is very simple, namely $(i/4\pi^2)(p-m)^{-1}$ that is

$$K_+(2, 1) = (i/4\pi^2)\int (p-m)^{-1} \exp(-ip\cdot x_{21})d^4p, \quad (31)$$

where $p\cdot x_{21} = p\cdot x_2 - p\cdot x_1 = p_\mu x_{2\mu} - p_\mu x_{1\mu}$, $p = p_\mu \gamma_\mu$, and d^4p means $(2\pi)^{-2}dp_1dp_2dp_3dp_4$, the integral over all p. That this is true can be seen immediately from (12), for the representation of the operator $i\nabla - m$ in energy (p_4) and momentum ($p_{1,2,3}$) space is $p-m$ and the transform of $\delta(2, 1)$ is a constant. The reciprocal matrix $(p-m)^{-1}$ can be interpreted as $(p+m)(p^2-m^2)^{-1}$ for $p^2-m^2=(p-m)(p+m)$ is a pure number not involving γ matrices. Hence if one wishes one can write

$$K_+(2, 1) = i(i\nabla_2 + m)I_+(2, 1),$$

where

$$I_+(2, 1) = (2\pi)^{-2}\int (p^2-m^2)^{-1} \exp(-ip\cdot x_{21})d^4p, \quad (32)$$

is not a matrix operator but a function satisfying

$$\Box_2{}^2 I_+(2, 1) - m^2 I_+(2, 1) = \delta(2, 1), \quad (33)$$

where $-\Box_2{}^2 = (\nabla_2)^2 = (\partial/\partial x_{2\mu})(\partial/\partial x_{2\mu})$.

The integrals (31) and (32) are not yet completely defined for there are poles in the integrand when $p^2-m^2=0$. We can define how these poles are to be evaluated by the rule that *m is considered to have an infinitesimal negative imaginary part*. That is m, is replaced by $m-i\delta$ and the limit taken as $\delta\to 0$ from above. This can be seen by imagining that we calculate K_+ by integrating on p_4 first. If we call $E = +(m^2+p_1^2+p_2^2+p_3^2)^{\frac{1}{2}}$ then the integrals involve p_4 essentially as $\int \exp(-ip_4(t_2-t_1))dp_4(p_4^2-E^2)^{-1}$ which has poles at $p_4=+E$ and $p_4=-E$. The replacement of m by $m-i\delta$ means that E has a small negative imaginary part; the first pole is below, the second above the real axis. Now if $t_2-t_1>0$ the contour can be completed around the semicircle below the real axis thus giving a residue from the $p_4=+E$ pole, or $-(2E)^{-1}\exp(-iE(t_2-t_1))$. If $t_2-t_1<0$ the upper semicircle must be used, and $p_4=-E$ at the pole, so that the function varies in each case as required by the other definition (17).

Other solutions of (12) result from other prescriptions. For example if p_4 in the factor $(p^2-m^2)^{-1}$ is considered to have a positive imaginary part K_+ becomes replaced by K_0, the Dirac one-electron kernel, zero for $t_2<t_1$. Explicitly the function is[11] ($\mathbf{x}$, $t=x_{21\mu}$)

$$I_+(\mathbf{x}, t) = -(4\pi)^{-1}\delta(s^2) + (m/8\pi s)H_1^{(2)}(ms), \quad (34)$$

where $s=+(t^2-\mathbf{x}^2)^{\frac{1}{2}}$ for $t^2>\mathbf{x}^2$ and $s=-i(x^2-t^2)^{\frac{1}{2}}$ for $t^2<x^2$, $H_1^{(2)}$ is the Hankel function and $\delta(s^2)$ is the Dirac delta function of s^2. It behaves asymptotically as $\exp(-ims)$, decaying exponentially in space-like directions.[12]

By means of such transforms the matrix elements like (22), (23) are easily worked out. A free particle wave function for an electron of momentum p_1 is $u_1\exp(-ip_1\cdot x)$ where u_1 is a constant spinor satisfying the Dirac equation $p_1u_1=mu_1$ so that $p_1^2=m^2$. The matrix element (22) for going from a state p_1, u_1 to a state of momentum p_2, spinor u_2, is $-4\pi^2 i(\bar{u}_2 a(q)u_1)$ where we have imagined A expanded in a Fourier integral

$$A(1) = \int a(q)\exp(-iq\cdot x_1)d^4q,$$

and we select the component of momentum $q=p_2-p_1$.

The second order term (23) is the matrix element between u_1 and u_2 of

$$-4\pi^2 i\int (a(p_2-p_1-q))(p_1+q-m)^{-1}a(q)d^4q, \quad (35)$$

since the electron of momentum p_1 may pick up q from the potential $a(q)$, propagate with momentum p_1+q (factor $(p_1+q-m)^{-1}$) until it is scattered again by the potential, $a(p_2-p_1-q)$, picking up the remaining momentum, p_2-p_1-q, to bring the total to p_2. Since all values of q are possible, one integrates over q.

These same matrices apply directly to positron problems, for if the time component of, say, p_1 is negative the state represents a positron of four-momentum $-p_1$, and we are describing pair production if p_2 is an electron, i.e., has positive time component, etc.

The probability of an event whose matrix element is $(\bar{u}_2Mu_1)$ is proportional to the absolute square. This may also be written $(\bar{u}_1\bar{M}u_2)(\bar{u}_2Mu_1)$, where $\bar{M}$ is M with the operators written in opposite order and explicit appearance of i changed to $-i$($\bar{M}$ is β times the complex conjugate transpose of βM). For many problems we are not concerned about the spin of the final state. Then we can sum the probability over the two u_2 corresponding to the two spin directions. This is not a complete set because p_2 has another eigenvalue, $-m$. To permit summing over all states we can insert the projection operator $(2m)^{-1}(p_2+m)$ and so obtain $(2m)^{-1}(\bar{u}_1\bar{M}(p_2+m)Mu_1)$ for the probability of transition from p_1, u_1, to p_2 with arbitrary spin. If the incident state is unpolarized we can sum on its spins too, and obtain

$$(2m)^{-2}Sp[(p_1+m)\bar{M}(p_2+m)M] \quad (36)$$

for (twice) the probability that an electron of arbitrary spin with momentum p_1 will make transition to p_2. The expressions are all valid for positrons when p's with

[11] $I_+(\mathbf{x}, t)$ is $(2i)^{-1}(D_1(\mathbf{x}, t) - iD(\mathbf{x}, t))$ where D_1 and D are the functions defined by W. Pauli, Rev. Mod. Phys. 13, 203 (1941).

[12] If the $-i\delta$ is kept with m here too the function I_+ approaches zero for infinite positive and negative times. This may be useful in general analyses in avoiding complications from infinitely remote surfaces.

negative energies are inserted, and the situation interpreted in accordance with the timing relations discussed above. (We have used functions normalized to $(\bar{u}u)=1$ instead of the conventional $(\bar{u}\beta u)=(u^*u)=1$. On our scale $(\bar{u}\beta u)=\text{energy}/m$ so the probabilities must be corrected by the appropriate factors.)

The author has many people to thank for fruitful conversations about this subject, particularly H. A. Bethe and F. J. Dyson.

APPENDIX

a. Deduction from Second Quantization

In this section we shall show the equivalence of this theory with the hole theory of the positron.[2] According to the theory of second quantization of the electron field in a given potential,[13] the state of this field at any time is represented by a wave function χ satisfying

$$i\partial\chi/\partial t=H\chi,$$

where $H=\int\Psi^*(\mathbf{x})(\boldsymbol{\alpha}\cdot(-i\nabla-\mathbf{A})+A_4+m\beta)\Psi(\mathbf{x})d^3\mathbf{x}$ and $\Psi(\mathbf{x})$ is an operator annihilating an electron at position $\mathbf{x}$, while $\Psi^*(\mathbf{x})$ is the corresponding creation operator. We contemplate a situation in which at $t=0$ we have present some electrons in states represented by ordinary spinor functions $f_1(\mathbf{x})$, $f_2(\mathbf{x})$, $\cdots$ assumed orthogonal, and some positrons. These are described as holes in the negative energy sea, the electrons which would normally fill the holes having wave functions $p_1(\mathbf{x})$, $p_2(\mathbf{x})$, $\cdots$. We ask, at time T what is the amplitude that we find electrons in states $g_1(\mathbf{x})$, $g_2(\mathbf{x})$, $\cdots$ and holes at $q_1(\mathbf{x})$, $q_2(\mathbf{x})$, $\cdots$. If the initial and final state vectors representing this situation are χ_i and χ_f respectively, we wish to calculate the matrix element

$$R=\left(\chi_f^*\exp\left(-i\int_0^T Hdt\right)\chi_i\right)=(\chi_f^*S\chi_i). \tag{37}$$

We assume that the potential A differs from zero only for times between 0 and T so that a vacuum can be defined at these times. If χ_0 represents the vacuum state (that is, all negative energy states filled, all positive energies empty), the amplitude for having a vacuum at time T, if we had one at $t=0$, is

$$C_v=(\chi_0^*S\chi_0), \tag{38}$$

writing S for $\exp(-i\int_0^T Hdt)$. Our problem is to evaluate R and show that it is a simple factor times C_v, and that the factor involves the $K_+^{(A)}$ functions in the way discussed in the previous sections.

To do this we first express χ_i in terms of χ_0. The operator

$$\Phi^*=\int\Psi^*(\mathbf{x})\phi(\mathbf{x})d^3\mathbf{x}, \tag{39}$$

creates an electron with wave function $\phi(\mathbf{x})$. Likewise $\Phi=\int\phi^*(\mathbf{x})\times\Psi(\mathbf{x})d^3\mathbf{x}$ annihilates one with wave function $\phi(\mathbf{x})$. Hence state χ_i is $\chi_i=F_1^*F_2^*\cdots P_1P_2\cdots\chi_0$ while the final state is $G_1^*G_2^*\cdots\times Q_1Q_2\cdots\chi_0$ where F_i, G_i, P_i, Q_i are operators defined like Φ, in (39), but with f_i, g_i, p_i, q_i replacing ϕ; for the initial state would result from the vacuum if we created the electrons in f_1, f_2, $\cdots$ and annihilated those in p_1, p_2, $\cdots$. Hence we must find

$$R=(\chi_0^*\cdots Q_2^*Q_1^*\cdots G_2G_1SF_1^*F_2^*\cdots P_1P_2\cdots\chi_0). \tag{40}$$

To simplify this we shall have to use commutation relations between a Φ^* operator and S. To this end consider $\exp(-i\int_0^t Hdt')\Phi^*\times\exp(+i\int_0^t Hdt')$ and expand this quantity in terms of $\Psi^*(\mathbf{x})$, giving $\int\Psi^*(\mathbf{x})\phi(\mathbf{x},t)d^3\mathbf{x}$, (which defines $\phi(\mathbf{x},t)$). Now multiply this equation by $\exp(+i\int_0^t Hdt')\cdots\exp(-i\int_0^t Hdt')$ and find

$$\int\Psi^*(\mathbf{x})\phi(\mathbf{x})d^3\mathbf{x}=\int\Psi^*(\mathbf{x},t)\phi(\mathbf{x},t)d^3\mathbf{x}, \tag{41}$$

where we have defined $\Psi(\mathbf{x},t)$ by $\Psi(\mathbf{x},t)=\exp(+i\int_0^t Hdt')\Psi(\mathbf{x})\times\exp(-i\int_0^t Hdt')$. As is well known $\Psi(\mathbf{x},t)$ satisfies the Dirac equation, (differentiate $\Psi(\mathbf{x},t)$ with respect to t and use commutation relations of H and Ψ)

$$i\partial\Psi(\mathbf{x},t)/\partial t=(\boldsymbol{\alpha}\cdot(-i\nabla-\mathbf{A})+A_4+m\beta)\Psi(\mathbf{x},t). \tag{42}$$

Consequently $\phi(\mathbf{x},t)$ must also satisfy the Dirac equation (differentiate (41) with respect to t, use (42) and integrate by parts).

That is, if $\phi(\mathbf{x},T)$ is that solution of the Dirac equation at time T which is $\phi(\mathbf{x})$ at $t=0$, and if we define $\Phi^*=\int\Psi^*(\mathbf{x})\phi(\mathbf{x})d^3\mathbf{x}$ and $\Phi'^*=\int\Psi^*(\mathbf{x})\phi(\mathbf{x},T)d^3\mathbf{x}$ then $\Phi'^*=S\Phi^*S^{-1}$, or

$$S\Phi^*=\Phi'^*S. \tag{43}$$

The principle on which the proof will be based can now be illustrated by a simple example. Suppose we have just one electron initially and finally and ask for

$$r=(\chi_0^*GSF^*\chi_0). \tag{44}$$

We might try putting F^* through the operator S using (43), $SF^*=F'^*S$, where f' in $F'^*=\int\Psi^*(\mathbf{x})f'(\mathbf{x})d^3\mathbf{x}$ is the wave function at T arising from $f(\mathbf{x})$ at 0. Then

$$r=(\chi_0^*GF'^*S\chi_0)=\int g^*(\mathbf{x})f'(\mathbf{x})d^3\mathbf{x}\cdot C_v-(\chi_0^*F'^*GS\chi_0), \tag{45}$$

where the second expression has been obtained by use of the definition (38) of C_v and the general commutation relation

$$GF^*+F^*G=\int g^*(\mathbf{x})f(\mathbf{x})d^3\mathbf{x},$$

which is a consequence of the properties of $\Psi(\mathbf{x})$ (the others are $FG=-GF$ and $F^*G^*=-G^*F^*$). Now $\chi_0^*F'^*$ in the last term in (45) is the complex conjugate of $F'\chi_0$. Thus if f' contained only positive energy components, $F'\chi_0$ would vanish and we would have reduced r to a factor times C_v. But F', as worked out here, does contain negative energy components created in the potential A and the method must be slightly modified.

Before putting F^* through the operator we shall add to it another operator F''^* arising from a function $f''(x)$ containing *only negative* energy components and so chosen that the resulting f' has *only positive* ones. That is we want

$$S(F_{\text{pos}}^*+F_{\text{neg}}''^*)=F_{\text{pos}}'^*S, \tag{46}$$

where the "pos" and "neg" serve as reminders of the sign of the energy components contained in the operators. This we can now use in the form

$$SF_{\text{pos}}^*=F_{\text{pos}}'^*S-SF_{\text{neg}}''^*. \tag{47}$$

In our one electron problem this substitution replaces r by two terms

$$r=(\chi_0^*GF_{\text{pos}}'^*S\chi_0)-(\chi_0^*GSF_{\text{neg}}''^*\chi_0).$$

The first of these reduces to

$$r=\int g^*(\mathbf{x})f_{\text{pos}}'(\mathbf{x})d^3\mathbf{x}\cdot C_v,$$

as above, for $F_{\text{pos}}'\chi_0$ is now zero, while the second is zero since the creation operator $F_{\text{neg}}''^*$ gives zero when acting on the vacuum state as all negative energies are full. This is the central idea of the demonstration.

The problem presented by (46) is this: Given a function $f_{\text{pos}}(\mathbf{x})$ at time 0, to find the amount, f_{neg}'', of negative energy component which must be added in order that the solution of Dirac's equation at time T will have only positive energy components, f_{pos}'. This is a boundary value problem for which the kernel $K_+^{(A)}$ is designed. We know the positive energy components initially, f_{pos}, and the negative ones finally (zero). The positive ones finally are therefore (using (19))

$$f_{\text{pos}}'(\mathbf{x}_2)=\int K_+^{(A)}(2,1)\beta f_{\text{pos}}(\mathbf{x}_1)d^3\mathbf{x}_1, \tag{48}$$

where $t_2=T$, $t_1=0$. Similarly, the negative ones initially are

$$f_{\text{neg}}''(\mathbf{x}_2)=\int K_+^{(A)}(2,1)\beta f_{\text{pos}}(\mathbf{x}_1)d^3\mathbf{x}_1-f_{\text{pos}}(\mathbf{x}_2), \tag{49}$$

where t_2 approaches zero from above, and $t_1=0$. The $f_{\text{pos}}(\mathbf{x}_2)$ is

[13] See, for example, G. Wentzel, *Einfuhrung in die Quantentheorie der Wellenfelder* (Franz Deuticke, Leipzig, 1943), Chapter V.

subtracted to keep in $f_{neg}''(\mathbf{x}_2)$ only those waves which return from the potential and not those arriving directly at t_2 from the $K_+(2,1)$ part of $K_+^{(A)}(2,1)$, as $t_2\rightarrow 0$. We could also have written

$$f_{neg}''(\mathbf{x}_2)=\int[K_+^{(A)}(2,1)-K_+(2,1)]\beta f_{pos}(\mathbf{x}_1)d^3\mathbf{x}_1. \quad (50)$$

Therefore the one-electron problem, $r=\int g^*(\mathbf{x})f_{pos}'(\mathbf{x})d^3\mathbf{x}\cdot C_v$, gives by (48)

$$r=C_v\int g^*(\mathbf{x}_2)K_+^{(A)}(2,1)\beta f(\mathbf{x}_1)d^3\mathbf{x}_1d^3\mathbf{x}_2,$$

as expected in accordance with the reasoning of the previous sections (i.e., (20) with $K_+^{(A)}$ replacing K_+).

The proof is readily extended to the more general expression R, (40), which can be analyzed by induction. First one replaces F_1^* by a relation such as (47) obtaining two terms

$$R=(\chi_0^*\cdots Q_2^*Q_1^*\cdots G_2G_1F_{1pos}'^*SF_2^*\cdots P_1P_2\cdots\chi_0)$$
$$-(\chi_0^*\cdots Q_2^*Q_1^*\cdots G_2G_1SF_{1neg}''^*F_2^*\cdots P_1P_2\cdots\chi_0).$$

In the first term the order of $F_{1pos}'^*$ and G_1 is then interchanged, producing an additional term $\int g_1^*(\mathbf{x})f_{1pos}'(\mathbf{x})d^3\mathbf{x}$ times an expression with one less electron in initial and final state. Next it is exchanged with G_2 producing an addition $-\int g_2^*(\mathbf{x})f_{1pos}'(\mathbf{x})d^3\mathbf{x}$ times a similar term, etc. Finally on reaching the Q_1^* with which it anticommutes it can be simply moved over to juxtaposition with χ_0^* where it gives zero. The second term is similarly handled by moving $F_{1neg}''^*$ through anti commuting F_2^*, etc., until it reaches P_1. Then it is exchanged with P_1 to produce an additional simpler term with a factor $\mp\int p_1^*(\mathbf{x})f_{1neg}''(\mathbf{x})d^3\mathbf{x}$ or $\mp\int p_1^*(\mathbf{x}_2)K_+^{(A)}(2,1)\beta f_1(\mathbf{x}_1)d^3\mathbf{x}_1d^3\mathbf{x}_2$ from (49), with $t_2=t_1=0$ (the extra $f_1(\mathbf{x}_2)$ in (49) gives zero as it is orthogonal to $p_1(\mathbf{x}_2)$). This describes in the expected manner the annihilation of the pair, electron f_1, positron p_1. The $F_{neg}''^*$ is moved in this way successively through the P's until it gives zero when acting on χ_0. Thus R is reduced, with the expected factors (and with alternating signs as required by the exclusion principle), to simpler terms containing two less operators which may in turn be further reduced by using F_2^* in a similar manner, etc. After all the F^* are used the Q^*'s can be reduced in a similar manner. They are moved through the S in the opposite direction in such a manner as to produce a purely negative energy operator at time 0, using relations analogous to (46) to (49). After all this is done we are left simply with the expected factor times C_v (assuming the net charge is the same in initial and final state.)

In this way we have written the solution to the general problem of the motion of electrons in given potentials. The factor C_v is obtained by normalization. However for photon fields it is desirable to have an explicit form for C_v in terms of the potentials. This is given by (30) and (29) and it is readily demonstrated that this also is correct according to second quantization.

b. Analysis of the Vacuum Problem

We shall calculate C_v from second quantization by induction considering a series of problems each containing a potential distribution more nearly like the one we wish. Suppose we know C_v for a problem like the one we want and having the same potentials for time t between some t_0 and T, but having potential zero for times from 0 to t_0. Call this $C_v(t_0)$, the corresponding Hamiltonian H_{t_0} and the sum of contributions for all single loops, $L(t_0)$. Then for $t_0=T$ we have zero potential at all times, no pairs can be produced, $L(T)=0$ and $C_v(T)=1$. For $t_0=0$ we have the complete problem, so that $C_v(0)$ is what is defined as C_v in (38). Generally we have,

$$C_v(t_0)=\left(\chi_0^*\exp\left(-i\int_0^T H_{t_0}dt\right)\chi_0\right)$$
$$=\left(\chi_0^*\exp\left(-i\int_{t_0}^T H_{t_0}dt\right)\chi_0\right),$$

since H_{t_0} is identical to the constant vacuum Hamiltonian H_T for $t<t_0$ and χ_0 is an eigenfunction of H_T with an eigenvalue (energy of vacuum) which we can take as zero.

The value of $C_v(t_0-\Delta t_0)$ arises from the Hamiltonian $H_{t_0-\Delta t_0}$ which differs from H_{t_0} just by having an extra potential during the short interval Δt_0. Hence, to first order in Δt_0, we have

$$C_v(t_0-\Delta t_0)=\left(\chi_0^*\exp\left(-i\int_{t_0-\Delta t_0}^T H_{t_0-\Delta t_0}dt\right)\chi_0\right)$$
$$=\left(\chi_0^*\exp\left(-i\int_{t_0}^T H_{t_0}dt\right)\left[1-i\Delta t_0\int\Psi^*(\mathbf{x})\right.\right.$$
$$\left.\left.\times(-\boldsymbol{\alpha}\cdot\mathbf{A}(\mathbf{x},t_0)+A_4(\mathbf{x},t_0))\Psi(\mathbf{x})d^3\mathbf{x}\right]\chi_0\right);$$

we therefore obtain for the derivative of C_v the expression

$$-dC_v(t_0)/dt_0=-i\left(\chi_0^*\exp\left(-i\int_{t_0}^T H_{t_0}dt\right)\right.$$
$$\left.\times\int\Psi^*(\mathbf{x})\beta A(\mathbf{x},t_0)\Psi(\mathbf{x})d^3\mathbf{x}\chi_0\right), \quad (51)$$

which will be reduced to a simple factor times $C_v(t_0)$ by methods analogous to those used in reducing R. The operator Ψ can be imagined to be split into two pieces Ψ_{pos} and Ψ_{neg} operating on positive and negative energy states respectively. The Ψ_{pos} on χ_0 gives zero so we are left with two terms in the current density, $\Psi_{pos}^*\beta A\Psi_{neg}$ and $\Psi_{neg}^*\beta A\Psi_{neg}$. The latter $\Psi_{neg}^*\beta A\Psi_{neg}$ is just the expectation value of βA taken over all negative energy states (minus $\Psi_{neg}\beta A\Psi_{neg}^*$ which gives zero acting on χ_0). This is the effect of the vacuum expectation current of the electrons in the sea which we should have subtracted from our original Hamiltonian in the customary way.

The remaining term $\Psi_{pos}^*\beta A\Psi_{neg}$, or its equivalent $\Psi_{pos}^*\beta A\Psi$ can be considered as $\Psi^*(\mathbf{x})\mathbf{f}_{pos}(\mathbf{x})$ where $\mathbf{f}_{pos}(\mathbf{x})$ is written for the positive energy component of the operator $\beta A\Psi(\mathbf{x})$. Now this operator, $\Psi^*(\mathbf{x})\mathbf{f}_{pos}(\mathbf{x})$, or more precisely just the $\Psi^*(\mathbf{x})$ part of it, can be pushed through the $\exp(-i\int_{t_0}^T Hdt)$ in a manner exactly analogous to (47) when f is a function. (An alternative derivation results from the consideration that the operator $\Psi(\mathbf{x},t)$ which satisfies the Dirac equation also satisfies the linear integral equations which are equivalent to it.) That is, (51) can be written by (48), (50),

$$-dC_v(t_0)/dt_0=-i\left(\chi_0^*\int\int\Psi^*(\mathbf{x}_2)K_+^{(A)}(2,1)\right.$$
$$\left.\times\exp\left(-i\int_{t_0}^T Hdt\right)A(1)\Psi(\mathbf{x}_1)d^3\mathbf{x}_1d^3\mathbf{x}_2\chi_0\right)$$
$$+i\left(\chi_0^*\exp\left(-i\int_{t_0}^T Hdt\right)\int\int\Psi^*(\mathbf{x}_2)[K_+^{(A)}(2,1)\right.$$
$$\left.-K_+(2,1)]A(1)\Psi(\mathbf{x}_1)d^3\mathbf{x}_1d^3\mathbf{x}_2\chi_0\right),$$

where in the first term $t_2=T$, and in the second $t_2\rightarrow t_0=t_1$. The (A) in $K_+^{(A)}$ refers to that part of the potential A after t_0. The first term vanishes for it involves (from the $K_+^{(A)}(2,1)$) only positive energy components of Ψ^*, which give zero operating into χ_0^*. In the second term only negative components of $\Psi^*(\mathbf{x}_2)$ appear. If, then $\Psi^*(\mathbf{x}_2)$ is interchanged in order with $\Psi(\mathbf{x}_1)$ it will give zero operating on χ_0, and only the term,

$$-dC_v(t_0)/dt_0=+i\int Sp[(K_+^{(A)}(1,1)$$
$$-K_+(1,1))A(1)]d^3\mathbf{x}_1\cdot C_v(t_0), \quad (52)$$

will remain, from the usual commutation relation of Ψ^* and Ψ.

The factor of $C_v(t_0)$ in (52) times $-\Delta t_0$ is, according to (29) (reference 10), just $L(t_0-\Delta t_0)-L(t_0)$ since this difference arises from the extra potential $\Delta A=A$ during the short time interval Δt_0. Hence $-dC_v(t_0)/dt_0=+(dL(t_0)/dt_0)C_v(t_0)$ so that integration from $t_0=T$ to $t_0=0$ establishes (30).

Starting from the theory of the electromagnetic field in second quantization, a deduction of the equations for quantum electrodynamics which appear in the succeeding paper may be worked out using very similar principles. The Pauli-Weisskopf theory of the Klein-Gordon equation can apparently be analyzed in essentially the same way as that used here for Dirac electrons.

PHYSICAL REVIEW VOLUME 76, NUMBER 6 SEPTEMBER 15, 1949

Space-Time Approach to Quantum Electrodynamics

R. P. FEYNMAN
Department of Physics, Cornell University, Ithaca, New York
(Received May 9, 1949)

In this paper two things are done. (1) It is shown that a considerable simplification can be attained in writing down matrix elements for complex processes in electrodynamics. Further, a physical point of view is available which permits them to be written down directly for any specific problem. Being simply a restatement of conventional electrodynamics, however, the matrix elements diverge for complex processes. (2) Electrodynamics is modified by altering the interaction of electrons at short distances. All matrix elements are now finite, with the exception of those relating to problems of vacuum polarization. The latter are evaluated in a manner suggested by Pauli and Bethe, which gives finite results for these matrices also. The only effects sensitive to the modification are changes in mass and charge of the electrons. Such changes could not be directly observed. Phenomena directly observable, are insensitive to the details of the modification used (except at extreme energies). For such phenomena, a limit can be taken as the range of the modification goes to zero. The results then agree with those of Schwinger. A complete, unambiguous, and presumably consistent, method is therefore available for the calculation of all processes involving electrons and photons.

The simplification in writing the expressions results from an emphasis on the over-all space-time view resulting from a study of the solution of the equations of electrodynamics. The relation of this to the more conventional Hamiltonian point of view is discussed. It would be very difficult to make the modification which is proposed if one insisted on having the equations in Hamiltonian form.

The methods apply as well to charges obeying the Klein-Gordon equation, and to the various meson theories of nuclear forces. Illustrative examples are given. Although a modification like that used in electrodynamics can make all matrices finite for all of the meson theories, for some of the theories it is no longer true that all directly observable phenomena are insensitive to the details of the modification used.

The actual evaluation of integrals appearing in the matrix elements may be facilitated, in the simpler cases, by methods described in the appendix.

THIS paper should be considered as a direct continuation of a preceding one[1] (I) in which the motion of electrons, neglecting interaction, was analyzed, by dealing directly with the *solution* of the Hamiltonian differential equations. Here the same technique is applied to include interactions and in that way to express in simple terms the solution of problems in quantum electrodynamics.

For most practical calculations in quantum electrodynamics the solution is ordinarily expressed in terms of a matrix element. The matrix is worked out as an expansion in powers of $e^2/\hbar c$, the successive terms corresponding to the inclusion of an increasing number of virtual quanta. It appears that a considerable simplification can be achieved in writing down these matrix elements for complex processes. Furthermore, each term in the expansion can be written down and understood directly from a physical point of view, similar to the space-time view in I. It is the purpose of this paper to describe how this may be done. We shall also discuss methods of handling the divergent integrals which appear in these matrix elements.

The simplification in the formulae results mainly from the fact that previous methods unnecessarily separated into individual terms processes that were closely related physically. For example, in the exchange of a quantum between two electrons there were two terms depending on which electron emitted and which absorbed the quantum. Yet, in the virtual states considered, timing relations are not significant. Olny the order of operators in the matrix must be maintained. We have seen (I), that in addition, processes in which virtual pairs are produced can be combined with others in which only positive energy electrons are involved. Further, the effects of longitudinal and transverse waves can be combined together. The separations previously made were on an unrelativistic basis (reflected in the circumstance that apparently momentum but not energy is conserved in intermediate states). When the terms are combined and simplified, the relativistic invariance of the result is self-evident.

We begin by discussing the solution in space and time of the Schrödinger equation for particles interacting instantaneously. The results are immediately generalizable to delayed interactions of relativistic electrons and we represent in that way the laws of quantum electrodynamics. We can then see how the matrix element for any process can be written down directly. In particular, the self-energy expression is written down.

So far, nothing has been done other than a restatement of conventional electrodynamics in other terms. Therefore, the self-energy diverges. A modification[2] in interaction between charges is next made, and it is shown that the self-energy is made convergent and corresponds to a correction to the electron mass. After the mass correction is made, other real processes are finite and insensitive to the "width" of the cut-off in the interaction.[3]

Unfortunately, the modification proposed is not completely satisfactory theoretically (it leads to some difficulties of conservation of energy). It does, however, seem consistent and satisfactory to define the matrix

[1] R. P. Feynman, Phys. Rev. 76, 749 (1949), hereafter called I.

[2] For a discussion of this modification in classical physics see R. P. Feynman, Phys. Rev. 74 939 (1948), hereafter referred to as A.

[3] A brief summary of the methods and results will be found in R. P. Feynman, Phys. Rev. 74, 1430 (1948), hereafter referred to as B.

element for all real processes as the limit of that computed here as the cut-off width goes to zero. A similar technique suggested by Pauli and by Bethe can be applied to problems of vacuum polarization (resulting in a renormalization of charge) but again a strict physical basis for the rules of convergence is not known.

After mass and charge renormalization, the limit of zero cut-off width can be taken for all real processes. The results are then equivalent to those of Schwinger[4] who does not make explicit use of the convergence factors. The method of Schwinger is to identify the terms corresponding to corrections in mass and charge and, previous to their evaluation, to remove them from the expressions for real processes. This has the advantage of showing that the results can be strictly independent of particular cut-off methods. On the other hand, many of the properties of the integrals are analyzed using formal properties of invariant propagation functions. But one of the properties is that the integrals are infinite and it is not clear to what extent this invalidates the demonstrations. A practical advantage of the present method is that ambiguities can be more easily resolved; simply by direct calculation of the otherwise divergent integrals. Nevertheless, it is not at all clear that the convergence factors do not upset the physical consistency of the theory. Although in the limit the two methods agree, neither method appears to be thoroughly satisfactory theoretically. Nevertheless, it does appear that we now have available a complete and definite method for the calculation of physical processes to any order in quantum electrodynamics.

Since we can write down the solution to any physical problem, we have a complete theory which could stand by itself. It will be theoretically incomplete, however, in two respects. First, although each term of increasing order in $e^2/\hbar c$ can be written down it would be desirable to see some way of expressing things in finite form to all orders in $e^2/\hbar c$ at once. Second, although it will be physically evident that the results obtained are equivalent to those obtained by conventional electrodynamics the mathematical proof of this is not included. Both of these limitations will be removed in a subsequent paper (see also Dyson[4]).

Briefly the genesis of this theory was this. The conventional electrodynamics was expressed in the Lagrangian form of quantum mechanics described in the Reviews of Modern Physics.[5] The motion of the field oscillators could be integrated out (as described in Section 13 of that paper), the result being an expression of the delayed interaction of the particles. Next the modification of the delta-function interaction could be made directly from the analogy to the classical case.[2] This was still not complete because the Lagrangian method had been worked out in detail only for particles obeying the non-relativistic Schrödinger equation. It was then modified in accordance with the requirements of the Dirac equation and the phenomenon of pair creation. This was made easier by the reinterpretation of the theory of holes (I). Finally for practical calculations the expressions were developed in a power series in $e^2/\hbar c$. It was apparent that each term in the series had a simple physical interpretation. Since the result was easier to understand than the derivation, it was thought best to publish the results first in this paper. Considerable time has been spent to make these first two papers as complete and as physically plausible as possible without relying on the Lagrangian method, because it is not generally familiar. It is realized that such a description cannot carry the conviction of truth which would accompany the derivation. On the other hand, in the interest of keeping simple things simple the derivation will appear in a separate paper.

The possible application of these methods to the various meson theories is discussed briefly. The formulas corresponding to a charge particle of zero spin moving in accordance with the Klein Gordon equation are also given. In an Appendix a method is given for calculating the integrals appearing in the matrix elements for the simpler processes.

The point of view which is taken here of the interaction of charges differs from the more usual point of view of field theory. Furthermore, the familiar Hamiltonian form of quantum mechanics must be compared to the over-all space-time view used here. The first section is, therefore, devoted to a discussion of the relations of these viewpoints.

1. COMPARISON WITH THE HAMILTONIAN METHOD

Electrodynamics can be looked upon in two equivalent and complementary ways. One is as the description of the behavior of a field (Maxwell's equations). The other is as a description of a direct interaction at a distance (albeit delayed in time) between charges (the solutions of Lienard and Wiechert). From the latter point of view light is considered as an interaction of the charges in the source with those in the absorber. This is an impractical point of view because many kinds of sources produce the same kind of effects. The field point of view separates these aspects into two simpler problems, production of light, and absorption of light. On the other hand, the field point of view is less practical when dealing with close collisions of particles (or their action on themselves). For here the source and absorber are not readily distinguishable, there is an intimate exchange of quanta. The fields are so closely determined by the motions of the particles that it is just as well not to separate the question into two problems but to consider the process as a direct interaction. Roughly, the field point of view is most practical for problems involv-

[4] J. Schwinger, Phys. Rev. 74, 1439 (1948), Phys. Rev. 75, 651 (1949). A proof of this equivalence is given by F. J. Dyson, Phys. Rev. 75, 486 (1949).

[5] R. P. Feynman, Rev. Mod. Phys. 20, 367 (1948). The application to electrodynamics is described in detail by H. J. Groenewold, Koninklijke Nederlandsche Akademia van Weteschappen. Proceedings Vol. LII, 3 (226) 1949.

ing real quanta, while the interaction view is best for the discussion of the virtual quanta involved. We shall emphasize the interaction viewpoint in this paper, first because it is less familiar and therefore requires more discussion, and second because the important aspect in the problems with which we shall deal is the effect of virtual quanta.

The Hamiltonian method is not well adapted to represent the direct action at a distance between charges because that action is delayed. The Hamiltonian method represents the future as developing out of the present. If the values of a complete set of quantities are known now, their values can be computed at the next instant in time. If particles interact through a delayed interaction, however, one cannot predict the future by simply knowing the present motion of the particles. One would also have to know what the motions of the particles were in the past in view of the interaction this may have on the future motions. This is done in the Hamiltonian electrodynamics, of course, by requiring that one specify besides the present motion of the particles, the values of a host of new variables (the coordinates of the field oscillators) to keep track of that aspect of the past motions of the particles which determines their future behavior. The use of the Hamiltonian forces one to choose the field viewpoint rather than the interaction viewpoint.

In many problems, for example, the close collisions of particles, we are not interested in the precise temporal sequence of events. It is not of interest to be able to say how the situation would look at each instant of time during a collision and how it progresses from instant to instant. Such ideas are only useful for events taking a long time and for which we can readily obtain information during the intervening period. For collisions it is much easier to treat the process as a whole.[6] The Møller interaction matrix for the the collision of two electrons is not essentially more complicated than the non-relativistic Rutherford formula, yet the mathematical machinery used to obtain the former from quantum electrodynamics is vastly more complicated than Schrödinger's equation with the e^2/r_{12} interaction needed to obtain the latter. The difference is only that in the latter the action is instantaneous so that the Hamiltonian method requires no extra variables, while in the former relativistic case it is delayed and the Hamiltonian method is very cumbersome.

We shall be discussing the solutions of equations rather than the time differential equations from which they come. We shall discover that the solutions, because of the over-all space-time view that they permit, are as easy to understand when interactions are delayed as when they are instantaneous.

As a further point, relativistic invariance will be self-evident. The Hamiltonian form of the equations develops the future from the instantaneous present. But for different observers in relative motion the instantaneous present is different, and corresponds to a different 3-dimensional cut of space-time. Thus the temporal analyses of different observers is different and their Hamiltonian equations are developing the process in different ways. These differences are irrelevant, however, for the solution is the same in any space time frame. By forsaking the Hamiltonian method, the wedding of relativity and quantum mechanics can be accomplished most naturally.

We illustrate these points in the next section by studying the solution of Schrödinger's equation for non-relativistic particles interacting by an instantaneous Coulomb potential (Eq. 2). When the solution is modified to include the effects of delay in the interaction and the relativistic properties of the electrons we obtain an expression of the laws of quantum electrodynamics (Eq. 4).

2. THE INTERACTION BETWEEN CHARGES

We study by the same methods as in I, the interaction of two particles using the same notation as I. We start by considering the non-relativistic case described by the Schrödinger equation (I, Eq. 1). The wave function at a given time is a function $\psi(\mathbf{x}_a, \mathbf{x}_b, t)$ of the coordinates $\mathbf{x}_a$ and $\mathbf{x}_b$ of each particle. Thus call $K(\mathbf{x}_a, \mathbf{x}_b, t; \mathbf{x}_a', \mathbf{x}_b', t')$ the amplitude that particle a at $\mathbf{x}_a'$ at time t' will get to $\mathbf{x}_a$ at t while particle b at $\mathbf{x}_b'$ at t' gets to $\mathbf{x}_b$ at t. If the particles are free and do not interact this is

$$K(\mathbf{x}_a, \mathbf{x}_b, t; \mathbf{x}_a', \mathbf{x}_b', t') = K_{0a}(\mathbf{x}_a, t; \mathbf{x}_a', t')K_{0b}(\mathbf{x}_b, t; \mathbf{x}_b', t')$$

where K_{0a} is the K_0 function for particle a considered as free. In *this* case we can obviously define a quantity like K, but for which the time t need not be the same for particles a and b (likewise for t'); e.g.,

$$K_0(3, 4; 1, 2) = K_{0a}(3, 1)K_{0b}(4, 2) \tag{1}$$

can be thought of as the amplitude that particle a goes from $\mathbf{x}_1$ at t_1 to $\mathbf{x}_3$ at t_3 and that particle b goes from $\mathbf{x}_2$ at t_2 to $\mathbf{x}_4$ at t_4.

When the particles do interact, one can only define the quantity $K(3, 4; 1, 2)$ precisely if the interaction vanishes between t_1 and t_2 and also between t_3 and t_4. In a real physical system such is not the case. There is such an enormous advantage, however, to the concept that we shall continue to use it, imagining that we can neglect the effect of interactions between t_1 and t_2 and between t_3 and t_4. For practical problems this means choosing such long time intervals $t_3 - t_1$ and $t_4 - t_2$ that the extra interactions near the end points have small relative effects. As an example, in a scattering problem it may well be that the particles are so well separated initially and finally that the interaction at these times is negligible. Again energy values can be defined by the average rate of change of phase over such long time intervals that errors initially and finally can be neglected. Inasmuch as any physical problem can be defined in terms of scattering processes we do not lose much in

[6] This is the viewpoint of the theory of the S matrix of Heisenberg.

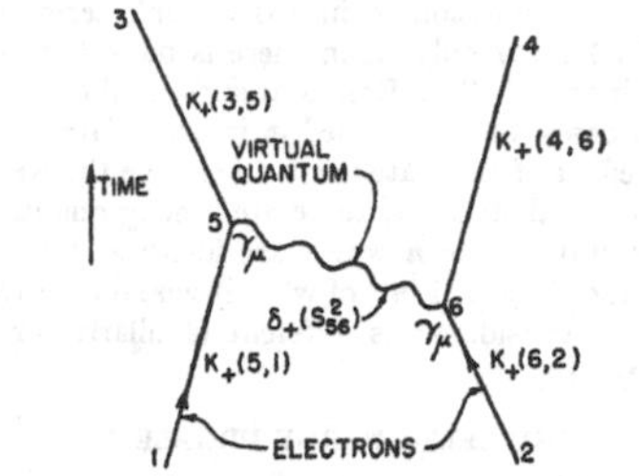

FIG. 1. The fundamental interaction Eq. (4). Exchange of one quantum between two electrons.

a general theoretical sense by this approximation. If it is not made it is not easy to study interacting particles relativistically, for there is nothing significant in choosing $t_1=t_3$ if $\mathbf{x}_1\neq\mathbf{x}_3$, as absolute simultaneity of events at a distance cannot be defined invariantly. It is essentially to avoid this approximation that the complicated structure of the older quantum electrodynamics has been built up. We wish to describe electrodynamics as a delayed interaction between particles. If we can make the approximation of assuming a meaning to $K(3,4;1,2)$ the results of this interaction can be expressed very simply.

To see how this may be done, imagine first that the interaction is simply that given by a Coulomb potential e^2/r where r is the distance between the particles. If this be turned on only for a very short time Δt_0 at time t_0, the first order correction to $K(3,4;1,2)$ can be worked out exactly as was Eq. (9) of I by an obvious generalization to two particles:

$$K^{(1)}(3,4;1,2)=-ie^2\int\int K_{0a}(3,5)K_{0b}(4,6)r_{56}^{-1}\times K_{0a}(5,1)K_{0b}(6,2)d^3\mathbf{x}_5d^3\mathbf{x}_6\Delta t_0,$$

where $t_5=t_6=t_0$. If now the potential were on at all times (so that strictly K is not defined unless $t_4=t_3$ and $t_1=t_2$), the first-order effect is obtained by integrating on t_0, which we can write as an integral over both t_5 and t_6 if we include a delta-function $\delta(t_5-t_6)$ to insure contribution only when $t_5=t_6$. Hence, the first-order effect of interaction is (calling $t_5-t_6=t_{56}$):

$$K^{(1)}(3,4;1,2)=-ie^2\int\int K_{0a}(3,5)K_{0b}(4,6)r_{56}^{-1}\times\delta(t_{56})K_{0a}(5,1)K_{0b}(6,2)d\tau_5d\tau_6, \quad (2)$$

where $d\tau=d^3\mathbf{x}dt$.

We know, however, in classical electrodynamics, that the Coulomb potential does not act instantaneously, but is delayed by a time r_{56}, taking the speed of light as unity. This suggests simply replacing $r_{56}^{-1}\delta(t_{56})$ in (2) by something like $r_{56}^{-1}\delta(t_{56}-r_{56})$ to represent the delay in the effect of b on a.

This turns out to be not quite right,[7] for when this interaction is represented by photons they must be of only positive energy, while the Fourier transform of $\delta(t_{56}-r_{56})$ contains frequencies of both signs. It should instead be replaced by $\delta_+(t_{56}-r_{56})$ where

$$\delta_+(x)=\int_0^\infty e^{-i\omega x}d\omega/\pi=\lim_{\epsilon\to 0}\frac{(\pi i)^{-1}}{x-i\epsilon}=\delta(x)+(\pi ix)^{-1}. \quad (3)$$

This is to be averaged with $r_{56}^{-1}\delta_+(-t_{56}-r_{56})$ which arises when $t_5<t_6$ and corresponds to a emitting the quantum which b receives. Since

$$(2r)^{-1}(\delta_+(t-r)+\delta_+(-t-r))=\delta_+(t^2-r^2),$$

this means $r_{56}^{-1}\delta(t_{56})$ is replaced by $\delta_+(s_{56}^2)$ where $s_{56}^2=t_{56}^2-r_{56}^2$ is the square of the relativistically invariant interval between points 5 and 6. Since in classical electrodynamics there is also an interaction through the vector potential, the complete interaction (see A, Eq. (1)) should be $(1-(\mathbf{v}_5\cdot\mathbf{v}_6)\delta_+(s_{56}^2)$, or in the relativistic case,

$$(1-\boldsymbol{\alpha}_a\cdot\boldsymbol{\alpha}_b)\delta_+(s_{56}^2)=\beta_a\beta_b\gamma_{a\mu}\gamma_{b\mu}\delta_+(s_{56}^2).$$

Hence we have for electrons obeying the Dirac equation,

$$K^{(1)}(3,4;1,2)=-ie^2\int\int K_{+a}(3,5)K_{+b}(4,6)\gamma_{a\mu}\gamma_{b\mu}\times\delta_+(s_{56}^2)K_{+a}(5,1)K_{+b}(6,2)d\tau_5d\tau_6, \quad (4)$$

where $\gamma_{a\mu}$ and $\gamma_{b\mu}$ are the Dirac matrices applying to the spinor corresponding to particles a and b, respectively (the factor $\beta_a\beta_b$ being absorbed in the definition, I Eq. (17), of K_+).

This is our fundamental equation for electrodynamics. It describes the effect of exchange of one quantum (therefore first order in e^2) between two electrons. It will serve as a prototype enabling us to write down the corresponding quantities involving the exchange of two or more quanta between two electrons or the interaction of an electron with itself. It is a consequence of conventional electrodynamics. Relativistic invariance is clear. Since one sums over μ it contains the effects of both longitudinal and transverse waves in a relativistically symmetrical way.

We shall now interpret Eq. (4) in a manner which will permit us to write down the higher order terms. It can be understood (see Fig. 1) as saying that the amplitude for "a" to go from 1 to 3 and "b" to go from 2 to 4 is altered to first order because they can exchange a quantum. Thus, "a" can go to 5 (amplitude $K_+(5,1)$)

[7] It, and a like term for the effect of a on b, leads to a theory which, in the classical limit, exhibits interaction through half-advanced and half-retarded potentials. Classically, this is equivalent to purely retarded effects within a closed box from which no light escapes (e.g., see A, or J. A. Wheeler and R. P. Feynman, Rev. Mod. Phys. 17, 157 (1945)). Analogous theorems exist in quantum mechanics but it would lead us too far astray to discuss them now.

emit a quantum (longitudinal, transverse, or scalar $\gamma_{a\mu}$) and then proceed to 3 ($K_+(3,5)$). Meantime "b" goes to 6 ($K_+(6,2)$), absorbs the quantum ($\gamma_{b\mu}$) and proceeds to 4 ($K_+(4,6)$). The quantum meanwhile proceeds from 5 to 6, which it does with amplitude $\delta_+(s_{56}{}^2)$. We must sum over all the possible quantum polarizations μ and positions and times of emission 5, and of absorption 6. Actually if $t_5 > t_6$ it would be better to say that "a" absorbs and "b" emits but no attention need be paid to these matters, as all such alternatives are automatically contained in (4).

The correct terms of higher order in e^2 or involving larger numbers of electrons (interacting with themselves or in pairs) can be written down by the same kind of reasoning. They will be illustrated by examples as we proceed. In a succeeding paper they will all be deduced from conventional quantum electrodynamics.

Calculation, from (4), of the transition element between positive energy free electron states gives the Möller scattering of two electrons, when account is taken of the Pauli principle.

The exclusion principle for interacting charges is handled in exactly the same way as for non-interacting charges (I). For example, for two charges it requires only that one calculate $K(3,4;1,2)-K(4,3;1,2)$ to get the net amplitude for arrival of charges at 3 and 4. It is disregarded in intermediate states. The interference effects for scattering of electrons by positrons discussed by Bhabha will be seen to result directly in this formulation. The formulas are interpreted to apply to positrons in the manner discussed in I.

As our primary concern will be for processes in which the quanta are virtual we shall not include here the detailed analysis of processes involving real quanta in initial or final state, and shall content ourselves by only stating the rules applying to them.[8] The result of the analysis is, as expected, that they can be included by the same line of reasoning as is used in discussing the virtual processes, provided the quantities are normalized in the usual manner to represent single quanta. For example, the amplitude that an electron in going from 1 to 2 absorbs a quantum whose vector potential, suitably normalized, is $c_\mu \exp(-ik\cdot x) = C_\mu(x)$ is just the expression (I, Eq. (13)) for scattering in a potential with $\mathbf{A}$ (3) replaced by $\mathbf{C}$ (3). Each quantum interacts only once (either in emission or in absorption), terms like (I, Eq. (14)) occur only when there is more than one quantum involved. The Bose statistics of the quanta can, in all cases, be disregarded in intermediate states. The only effect of the statistics is to change the weight of initial or final states. If there are among quanta, in the initial state, some n which are identical then the weight of the state is $(1/n!)$ of what it would be if these quanta were considered as different (similarly for the final state).

[8] Although in the expressions stemming from (4) the quanta are virtual, this is not actually a theoretical limitation. One way to deduce the correct rules for real quanta from (4) is to note that in a closed system all quanta can be considered as virtual (i.e., they have a known source and are eventually absorbed) so that in such a system the present description is complete and equivalent to the conventional one. In particular, the relation of the Einstein A and B coefficients can be deduced. A more practical direct deduction of the expressions for real quanta will be given in the subsequent paper. It might be noted that (4) can be rewritten as describing the action on a, $K^{(1)}(3,1) = i\int K_+(3,5) \times A(5) K_+(5,1) d\tau_5$ of the potential $A_\mu(5) = e^2 \int K_+(4,6)\delta_+(s_{56}{}^2)\gamma_\mu \times K_+(6,2) d\tau_6$ arising from Maxwell's equations $-\square^2 A_\mu = 4\pi j_\mu$ from a "current" $j_\mu(6) = e^2 K_+(4,6)\gamma_\mu K_+(6,2)$ produced by particle b in going from 2 to 4. This is virtue of the fact that δ_+ satisfies

$$-\square_2{}^2 \delta_+(s_{21}{}^2) = 4\pi\delta(2,1). \qquad (5)$$

3. THE SELF-ENERGY PROBLEM

Having a term representing the mutual interaction of a pair of charges, we must include similar terms to represent the interaction of a charge with itself. For under some circumstances what appears to be two distinct electrons may, according to I, be viewed also as a single electron (namely in case one electron was created in a pair with a positron destined to annihilate the other electron). Thus to the interaction between such electrons must correspond the possibility of the action of an electron on itself.[9]

This interaction is the heart of the self energy problem. Consider to first order in e^2 the action of an electron on itself in an otherwise force free region. The amplitude $K(2,1)$ for a single particle to get from 1 to 2 differs from $K_+(2,1)$ to first order in e^2 by a term

$$K^{(1)}(2,1) = -ie^2 \int\int K_+(2,4)\gamma_\mu K_+(4,3)\gamma_\mu \times K_+(3,1) d\tau_3 d\tau_4 \delta_+(s_{43}{}^2). \qquad (6)$$

It arises because the electron instead of going from 1 directly to 2, may go (Fig. 2) first to 3, ($K_+(3,1)$), emit a quantum (γ_μ), proceed to 4, ($K_+(4,3)$), absorb it (γ_μ), and finally arrive at 2 ($K_+(2,4)$). The quantum must go from 3 to 4 ($\delta_+(s_{43}{}^2)$).

This is related to the self-energy of a free electron in the following manner. Suppose initially, time t_1, we have an electron in state $f(1)$ which we imagine to be a positive energy solution of Dirac's equation for a free particle. After a long time $t_2 - t_1$ the perturbation will alter

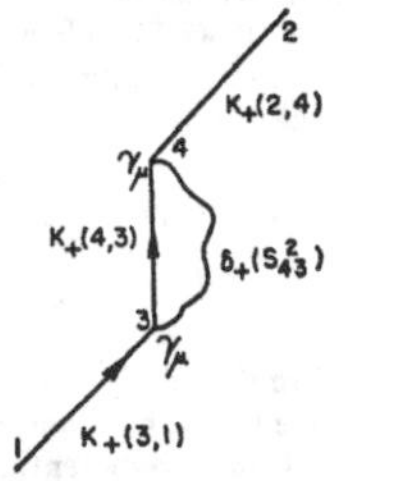

FIG. 2. Interaction of an electron with itself, Eq. (6).

[9] These considerations make it appear unlikely that the contention of J. A. Wheeler and R. P. Feynman, Rev. Mod. Phys. 17, 157 (1945), that electrons do not act on themselves, will be a successful concept in quantum electrodynamics.

the wave function, which can then be looked upon as a superposition of free particle solutions (actually it only contains f). The amplitude that $g(2)$ is contained is calculated as in (I, Eq. (21)). The diagonal element ($g=f$) is therefore

$$\int\int \bar{f}(2)\beta K^{(1)}(2,1)\beta f(1)d^3\mathbf{x}_1 d^3\mathbf{x}_2. \qquad (7)$$

The time interval $T=t_2-t_1$ (and the spatial volume V over which one integrates) must be taken very large, for the expressions are only approximate (analogous to the situation for two interacting charges).[10] This is because, for example, we are dealing incorrectly with quanta emitted just before t_2 which would normally be reabsorbed at times after t_2.

If $K^{(1)}(2,1)$ from (6) is actually substituted into (7) the surface integrals can be performed as was done in obtaining I, Eq. (22) resulting in

$$-ie^2\int\int \bar{f}(4)\gamma_\mu K_+(4,3)\gamma_\mu f(3)\delta_+(s_{43}^2)d\tau_3 d\tau_4. \qquad (8)$$

Putting for $f(1)$ the plane wave $u\exp(-ip\cdot x_1)$ where p_μ is the energy (p_4) and momentum of the electron ($p^2=m^2$), and u is a constant 4-index symbol, (8) becomes

$$-ie^2\int\int(\bar{u}\gamma_\mu K_+(4,3)\gamma_\mu u)$$
$$\times\exp(ip\cdot(x_4-x_3))\delta_+(s_{43}^2)d\tau_3 d\tau_4,$$

the integrals extending over the volume V and time interval T. Since $K_+(4,3)$ depends only on the difference of the coordinates of 4 and 3, $x_{43\mu}$, the integral on 4 gives a result (except near the surfaces of the region) independent of 3. When integrated on 3, therefore, the result is of order VT. The effect is proportional to V, for the wave functions have been normalized to unit volume. If normalized to volume V, the result would simply be proportional to T. This is expected, for if the effect were equivalent to a change in energy ΔE, the amplitude for arrival in f at t_2 is altered by a factor $\exp(-i\Delta E(t_2-t_1))$, or to first order by the difference $-i(\Delta E)T$. Hence, we have

$$\Delta E=e^2\int(\bar{u}\gamma_\mu K_+(4,3)\gamma_\mu u)\exp(ip\cdot x_{43})\delta_+(s_{43}^2)d\tau_4, \qquad (9)$$

integrated over all space-time $d\tau_4$. This expression will be simplified presently. In interpreting (9) we have tacitly assumed that the wave functions are normalized so that $(u^*u)=(\bar{u}\gamma_4 u)=1$. The equation may therefore be made independent of the normalization by writing the left side as $(\Delta E)(\bar{u}\gamma_4 u)$, or since $(\bar{u}\gamma_4 u)=(E/m)(\bar{u}u)$ and $m\Delta m=E\Delta E$, as $\Delta m(\bar{u}u)$ where Δm is an equivalent change in mass of the electron. In this form invariance is obvious.

One can likewise obtain an expression for the energy shift for an electron in a hydrogen atom. Simply replace K_+ in (8), by $K_+^{(V)}$, the exact kernel for an electron in the potential, $V=\beta e^2/r$, of the atom, and f by a wave function (of space and time) for an atomic state. In general the ΔE which results is not real. The imaginary part is negative and in $\exp(-i\Delta ET)$ produces an exponentially decreasing amplitude with time. This is because we are asking for the amplitude that an atom initially with no photon in the field, will still appear after time T with no photon. If the atom is in a state which can radiate, this amplitude must decay with time. The imaginary part of ΔE when calculated does indeed give the correct rate of radiation from atomic states. It is zero for the ground state and for a free electron.

In the non-relativistic region the expression for ΔE can be worked out as has been done by Bethe.[11] In the relativistic region (points 4 and 3 as close together as a Compton wave-length) the $K_+^{(V)}$ which should appear in (8) can be replaced to first order in V by K_+ plus $K_+^{(1)}(2,1)$ given in I, Eq. (13). The problem is then very similar to the radiationless scattering problem discussed below.

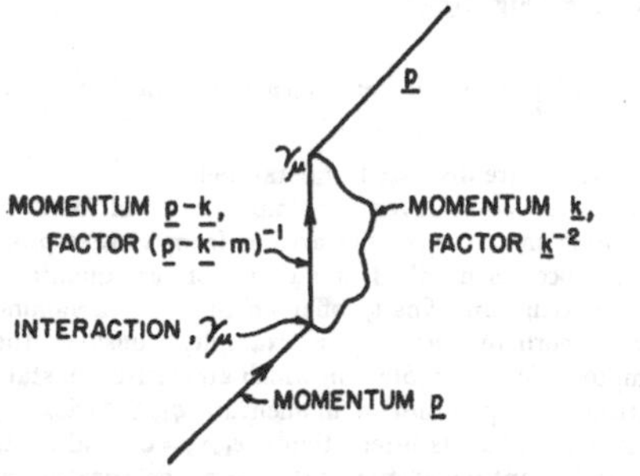

FIG. 3. Interaction of an electron with itself. Momentum space, Eq. (11).

4. EXPRESSION IN MOMENTUM AND ENERGY SPACE

The evaluation of (9), as well as all the other more complicated expressions arising in these problems, is very much simplified by working in the momentum and energy variables, rather than space and time. For this we shall need the Fourier Transform of $\delta_+(s_{21}^2)$ which is

$$-\delta_+(s_{21}^2)=\pi^{-1}\int\exp(-ik\cdot x_{21})k^{-2}d^4k, \qquad (10)$$

which can be obtained from (3) and (5) or from I, Eq. (32) noting that $I_+(2,1)$ for $m^2=0$ is $\delta_+(s_{21}^2)$ from

[10] This is discussed in reference 5 in which it is pointed out that the concept of a wave function loses accuracy if there are delayed self-actions.

[11] H. A. Bethe, Phys. Rev. 72, 339 (1947).

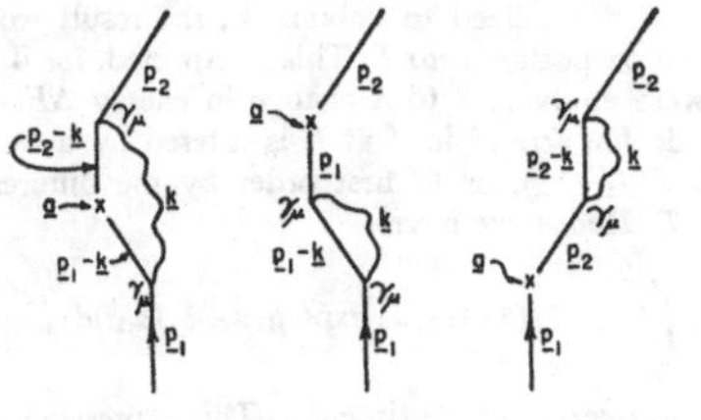

a. Eq. 12 b. Eq. 13 c. Eq. 14

FIG. 4. Radiative correction to scattering, momentum space.

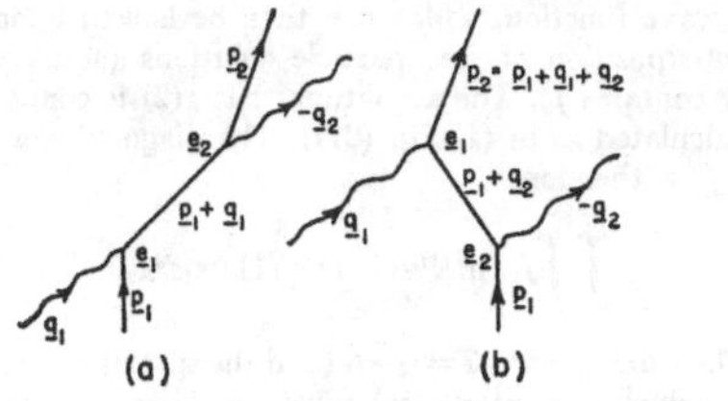

FIG. 5. Compton scattering, Eq. (15).

I, Eq. (34). The k^{-2} means $(k\cdot k)^{-1}$ or more precisely the limit as $\delta\to 0$ of $(k\cdot k+i\delta)^{-1}$. Further d^4k means $(2\pi)^{-2}dk_1dk_2dk_3dk_4$. If we imagine that quanta are particles of zero mass, then we can make the general rule that all poles are to be resolved by considering the masses of the particles and quanta to have infinitesimal negative imaginary parts.

Using these results we see that the self-energy (9) is the matrix element between $\bar{u}$ and u of the matrix

$$(e^2/\pi i)\int \gamma_\mu(p-k-m)^{-1}\gamma_\mu k^{-2}d^4k, \quad (11)$$

where we have used the expression (I, Eq. (31)) for the Fourier transform of K_+. This form for the self-energy is easier to work with than is (9).

The equation can be understood by imagining (Fig. 3) that the electron of momentum p emits (γ_μ) a quantum of momentum k, and makes its way now with momentum $p-k$ to the next event (factor $(p-k-m)^{-1}$) which is to absorb the quantum (another γ_μ). The amplitude of propagation of quanta is k^{-2}. (There is a factor $e^2/\pi i$ for each virtual quantum). One integrates over all quanta. The reason an electron of momentum p propagates as $1/(p-m)$ is that this operator is the reciprocal of the Dirac equation operator, and we are simply solving this equation. Likewise light goes as $1/k^2$, for this is the reciprocal D'Alembertian operator of the wave equation of light. The first γ_μ represents the current which generates the vector potential, while the second is the velocity operator by which this potential is multiplied in the Dirac equation when an external field acts on an electron.

Using the same line of reasoning, other problems may be set up directly in momentum space. For example, consider the scattering in a potential $A=A_\mu\gamma_\mu$ varying in space and time as $a\exp(-iq\cdot x)$. An electron initially in state of momentum $p_1=p_{1\mu}\gamma_\mu$ will be deflected to state p_2 where $p_2=p_1+q$. The zero-order answer is simply the matrix element of a between states 1 and 2. We next ask for the first order (in e^2) radiative correction due to virtual radiation of one quantum. There are several ways this can happen. First for the case illustrated in Fig. 4(a), find the matrix:

$$(e^2/\pi i)\int \gamma_\mu(p_2-k-m)^{-1}a(p_1-k-m)^{-1}\gamma_\mu k^{-2}d^4k. \quad (12)$$

For in this case, first[12] a quantum of momentum k is emitted (γ_μ), the electron then having momentum p_1-k and hence propagating with factor $(p_1-k-m)^{-1}$. Next it is scattered by the potential (matrix a) receiving additional momentum q, propagating on then (factor $(p_2-k-m)^{-1}$) with the new momentum until the quantum is reabsorbed (γ_μ). The quantum propagates from emission to absorption (k^{-2}) and we integrate over all quanta (d^4k), and sum on polarization μ. When this is integrated on k_4, the result can be shown to be exactly equal to the expressions (16) and (17) given in B for the same process, the various terms coming from residues of the poles of the integrand (12).

Or again if the quantum is both emitted and reabsorbed before the scattering takes place one finds (Fig. 4(b))

$$(e^2/\pi i)\int a(p_1-m)^{-1}\gamma_\mu(p_1-k-m)^{-1}\gamma_\mu k^{-2}d^4k, \quad (13)$$

or if both emission and absorption occur after the scattering, (Fig. 4(c))

$$(e^2/\pi i)\int \gamma_\mu(p_2-k-m)^{-1}\gamma_\mu(p_2-m)^{-1}ak^{-2}d^4k. \quad (14)$$

These terms are discussed in detail below.

We have now achieved our simplification of the form of writing matrix elements arising from virtual processes. Processes in which a number of real quanta is given initially and finally offer no problem (assuming correct normalization). For example, consider the Compton effect (Fig. 5(a)) in which an electron in state p_1 absorbs a quantum of momentum q_1, polarization vector $e_{1\mu}$ so that its interaction is $e_{1\mu}\gamma_\mu=e_1$, and emits a second quantum of momentum $-q_2$, polarization e_2 to arrive in final state of momentum p_2. The matrix for

[12] First, next, etc., here refer not to the order in true time but to the succession of events along the trajectory of the electron. That is, more precisely, to the order of appearance of the matrices in the expressions.

this process is $e_2(p_1+q_1-m)^{-1}e_1$. The total matrix for the Compton effect is, then,

$$e_2(p_1+q_1-m)^{-1}e_1+e_1(p_1+q_2-m)^{-1}e_2, \quad (15)$$

the second term arising because the emission of e_2 may also precede the absorption of e_1 (Fig. 5(b)). One takes matrix elements of this between initial and final electron states $(p_1+q_1=p_2-q_2)$, to obtain the Klein Nishina formula. Pair annihilation with emission of two quanta, etc., are given by the same matrix, positron states being those with negative time component of p. Whether quanta are absorbed or emitted depends on whether the time component of q is positive or negative.

5. THE CONVERGENCE OF PROCESSES WITH VIRTUAL QUANTA

These expressions are, as has been indicated, no more than a re-expression of conventional quantum electrodynamics. As a consequence, many of them are meaningless. For example, the self-energy expression (9) or (11) gives an infinite result when evaluated. The infinity arises, apparently, from the coincidence of the δ-function singularities in $K_+(4, 3)$ and $\delta_+(s_{43}^2)$. Only at this point is it necessary to make a real departure from conventional electrodynamics, a departure other than simply rewriting expressions in a simpler form.

We desire to make a modification of quantum electrodynamics analogous to the modification of classical electrodynamics described in a previous article, A. There the $\delta(s_{12}^2)$ appearing in the action of interaction was replaced by $f(s_{12}^2)$ where $f(x)$ is a function of small width and great height.

The obvious corresponding modification in the quantum theory is to replace the $\delta_+(s^2)$ appearing the quantum mechanical interaction by a new function $f_+(s^2)$. We can postulate that if the Fourier transform of the classical $f(s_{12}^2)$ is the integral over all k of $F(k^2)\exp(-ik\cdot x_{12})d^4k$, then the Fourier transform of $f_+(s^2)$ is the same integral taken over only positive frequencies k_4 for $t_2>t_1$ and over only negative ones for $t_2<t_1$ in analogy to the relation of $\delta_+(s^2)$ to $\delta(s^2)$. The function $f(s^2)=f(x\cdot x)$ can be written* as

$$f(x\cdot x)=(2\pi)^{-2}\int_{k_4=0}^{\infty}\int \sin(k_4|x_4|)\times\cos(\mathbf{K}\cdot\mathbf{x})dk_4d^3\mathbf{K}g(k\cdot k),$$

where $g(k\cdot k)$ is k_4^{-1} times the density of oscillators and may be expressed for positive k_4 as (A, Eq. (16))

$$g(k^2)=\int_0^{\infty}(\delta(k^2)-\delta(k^2-\lambda^2))G(\lambda)d\lambda,$$

where $\int_0^\infty G(\lambda)d\lambda=1$ and G involves values of λ large compared to m. This simply means that the amplitude for propagation of quanta of momentum k is

$$-F_+(k^2)=\pi^{-1}\int_0^{\infty}(k^{-2}-(k^2-\lambda^2)^{-1})G(\lambda)d\lambda,$$

rather than k^{-2}. That is, writing $F_+(k^2)=-\pi^{-1}k^{-2}C(k^2)$,

$$-f_+(s_{12}^2)=\pi^{-1}\int \exp(-ik\cdot x_{12})k^{-2}C(k^2)d^4k. \quad (16)$$

Every integral over an intermediate quantum which previously involved a factor d^4k/k^2 is now supplied with a convergence factor $C(k^2)$ where

$$C(k^2)=\int_0^{\infty}-\lambda^2(k^2-\lambda^2)^{-1}G(\lambda)d\lambda. \quad (17)$$

The poles are defined by replacing k^2 by $k^2+i\delta$ in the limit $\delta\to 0$. That is λ^2 may be assumed to have an infinitesimal negative imaginary part.

The function $f_+(s_{12}^2)$ may still have a discontinuity in value on the light cone. This is of no influence for the Dirac electron. For a particle satisfying the Klein Gordon equation, however, the interaction involves gradients of the potential which reinstates the δ function if f has discontinuities. The condition that f is to have no discontinuity in value on the light cone implies $k^2C(k^2)$ approaches zero as k^2 approaches infinity. In terms of $G(\lambda)$ the condition is

$$\int_0^{\infty}\lambda^2G(\lambda)d\lambda=0. \quad (18)$$

This condition will also be used in discussing the convergence of vacuum polarization integrals.

The expression for the self-energy matrix is now

$$(e^2/\pi i)\int \gamma_\mu(p-k-m)^{-1}\gamma_\mu k^{-2}d^4kC(k^2), \quad (19)$$

which, since $C(k^2)$ falls off at least as rapidly as $1/k^2$, converges. For practical purposes we shall suppose hereafter that $C(k^2)$ is simply $-\lambda^2/(k^2-\lambda^2)$ implying that some average (with weight $G(\lambda)d\lambda$) over values of λ may be taken afterwards. Since in all processes the quantum momentum will be contained in at least one extra factor of the form $(p-k-m)^{-1}$ representing propagation of an electron while that quantum is in the field, we can expect all such integrals with their convergence factors to converge and that the result of all such processes will now be finite and definite (excepting the processes with closed loops, discussed below, in which the diverging integrals are over the momenta of the electrons rather than the quanta).

The integral of (19) with $C(k^2)=-\lambda^2(k^2-\lambda^2)^{-1}$ noting that $p^2=m^2$, $\lambda\gg m$ and dropping terms of order m/λ, is (see Appendix A)

$$(e^2/2\pi)[4m(\ln(\lambda/m)+\tfrac{1}{2})-p(\ln(\lambda/m)+5/4)]. \quad (20)$$

* This relation is given incorrectly in A, equation just preceding 16.

When applied to a state of an electron of momentum p satisfying $pu=mu$, it gives for the change in mass (as in B, Eq. (9))

$$\Delta m = m(e^2/2\pi)(3\ln(\lambda/m)+\tfrac{3}{4}). \qquad (21)$$

6. RADIATIVE CORRECTIONS TO SCATTERING

We can now complete the discussion of the radiative corrections to scattering. In the integrals we include the convergence factor $C(k^2)$, so that they converge for large k. Integral (12) is also not convergent because of the well-known infra-red catastrophy. For this reason we calculate (as discussed in B) the value of the integral assuming the photons to have a small mass $\lambda_{\min}\ll m\ll\lambda$. The integral (12) becomes

$$(e^2/\pi i)\int \gamma_\mu(p_2-k-m)^{-1}a(p_1-k-m)^{-1} \times\gamma_\mu(k^2-\lambda_{\min}^2)^{-1}d^4kC(k^2-\lambda_{\min}^2),$$

which when integrated (see Appendix B) gives $(e^2/2\pi)$ times

$$\left[2\left(\ln\frac{m}{\lambda_{\min}}-1\right)\left(1-\frac{2\theta}{\tan 2\theta}\right)+\theta\tan\theta + \frac{4}{\tan 2\theta}\int_0^\theta \alpha\tan\alpha d\alpha\right]a + \frac{1}{4m}(qa-aq)\frac{2\theta}{\sin 2\theta}+ra, \qquad (22)$$

where $(q^2)^{\frac{1}{2}}=2m\sin\theta$ and we have assumed the matrix to operate between states of momentum p_1 and $p_2=p_1+q$ and have neglected terms of order $\lambda_{\min}/m$, m/λ, and q^2/λ^2. Here the only dependence on the convergence factor is in the term ra, where

$$r=\ln(\lambda/m)+9/4-2\ln(m/\lambda_{\min}). \qquad (23)$$

As we shall see in a moment, the other terms (13), (14) give contributions which just cancel the ra term. The remaining terms give for small q,

$$(e^2/4\pi)\left(\frac{1}{2m}(qa-aq)+\frac{4q^2}{3m^2}a\left(\ln\frac{m}{\lambda_{\min}}-\frac{3}{8}\right)\right), \qquad (24)$$

which shows the change in magnetic moment and the Lamb shift as interpreted in more detail in B.[13]

We must now study the remaining terms (13) and (14). The integral on k in (13) can be performed (after multiplication by $C(k^2)$) since it involves nothing but the integral (19) for the self-energy and the result is allowed to operate on the initial state u_1, (so that $p_1u_1=mu_1$). Hence the factor following $a(p_1-m)^{-1}$ will be just Δm. But, if one now tries to expand $1/(p_1-m) = (p_1+m)/(p_1^2-m^2)$ one obtains an infinite result, since $p_1^2=m^2$. This is, however, just what is expected physically. For the quantum can be emitted and absorbed at any time previous to the scattering. Such a process has the effect of a change in mass of the electron in the state 1. It therefore changes the energy by ΔE and the amplitude to first order in ΔE by $-i\Delta E\cdot t$ where t is the time it is acting, which is infinite. That is, the major effect of this term would be canceled by the effect of change of mass Δm.

The situation can be analyzed in the following manner. We suppose that the electron approaching the scattering potential a has not been free for an infinite time, but at some time far past suffered a scattering by a potential b. If we limit our discussion to the effects of Δm and of the virtual radiation of one quantum between two such scatterings each of the effects will be finite, though large, and their difference is determinate. The propagation from b to a is represented by a matrix

$$a(p'-m)^{-1}b, \qquad (25)$$

in which one is to integrate possibly over p' (depending on details of the situation). (If the time is long between b and a, the energy is very nearly determined so that p'^2 is very nearly m^2.)

We shall compare the effect on the matrix (25) of the virtual quanta and of the change of mass Δm. The effect of a virtual quantum is

$$(e^2/\pi i)\int a(p'-m)^{-1}\gamma_\mu(p'-k-m)^{-1} \times\gamma_\mu(p'-m)^{-1}bk^{-2}d^4kC(k^2), \qquad (26)$$

while that of a change of mass can be written

$$a(p'-m)^{-1}\Delta m(p'-m)^{-1}b, \qquad (27)$$

and we are interested in the difference (26)-(27). A simple and direct method of making this comparison is just to evaluate the integral on k in (26) and subtract from the result the expression (27) where Δm is given in (21). The remainder can be expressed as a multiple $-r(p'^2)$ of the unperturbed amplitude (25);

$$-r(p'^2)a(p'-m)^{-1}b. \qquad (28)$$

This has the same result (to this order) as replacing the potentials a and b in (25) by $(1-\frac{1}{2}r(p'^2))a$ and

[13] That the result given in B in Eq. (19) was in error was repeatedly pointed out to the author, in private communication, by V. F. Weisskopf and J. B. French, as their calculation, completed simultaneously with the author's early in 1948, gave a different result. French has finally shown that although the expression for the radiationless scattering B, Eq. (18) or (24) above is correct, it was incorrectly joined onto Bethe's non-relativistic result. He shows that the relation $\ln 2k_{\max}-1=\ln\lambda_{\min}$ used by the author should have been $\ln 2k_{\max}-5/6=\ln\lambda_{\min}$. This results in adding a term $-(1/6)$ to the logarithm in B, Eq. (19) so that the result now agrees with that of J. B. French and V. F. Weisskopf, Phys. Rev. 75, 1240 (1949) and N. H. Kroll and W. E. Lamb, Phys. Rev. 75, 388 (1949). The author feels unhappily responsible for the very considerable delay in the publication of French's result occasioned by this error. This footnote is appropriately numbered.

$(1-\frac{1}{2}r(p'^2))b$. In the limit, then, as $p'^2 \to m^2$ the net effect on the scattering is $-\frac{1}{2}ra$ where r, the limit of $r(p'^2)$ as $p'^2 \to m^2$ (assuming the integrals have an infra-red cut-off), turns out to be just equal to that given in (23). An equal term $-\frac{1}{2}ra$ arises from virtual transitions after the scattering (14) so that the entire ra term in (22) is canceled.

The reason that r is just the value of (12) when $q^2=0$ can also be seen without a direct calculation as follows: Let us call p the vector of length m in the direction of p' so that if $p'^2=m(1+\epsilon)^2$ we have $p'=(1+\epsilon)p$ and we take ϵ as very small, being of order T^{-1} where T is the time between the scatterings b and a. Since $(p'-m)^{-1}=(p'+m)/(p'^2-m^2)\approx(p+m)/2m^2\epsilon$, the quantity (25) is of order ϵ^{-1} or T. We shall compute corrections to it only to its own order (ϵ^{-1}) in the limit $\epsilon \to 0$. The term (27) can be written approximately[14] as

$$(e^2/\pi i)\int a(p'-m)^{-1}\gamma_\mu(p-k-m)^{-1} \times\gamma_\mu(p'-m)^{-1}bk^{-2}d^4kC(k^2),$$

using the expression (19) for Δm. The net of the two effects is therefore approximately[15]

$$-(e^2/\pi i)\int a(p'-m)^{-1}\gamma_\mu(p-k-m)^{-1}\epsilon p(p-k-m)^{-1} \times\gamma_\mu(p'-m)^{-1}bk^{-2}d^4kC(k^2),$$

a term now of order $1/\epsilon$ (since $(p'-m)^{-1}\approx(p+m)\times(2m^2\epsilon)^{-1}$) and therefore the one desired in the limit. Comparison to (28) gives for r the expression

$$(p_1+m/2m)\int\gamma_\mu(p_1-k-m)^{-1}(p_1m^{-1})(p_1-k-m)^{-1} \times\gamma_\mu k^{-2}d^4kC(k^2). \quad (29)$$

The integral can be immediately evaluated, since it is the same as the integral (12), but with $q=0$, for a replaced by p_1/m. The result is therefore $r\cdot(p_1/m)$ which when acting on the state u_1 is just r, as $p_1u_1=mu_1$. For the same reason the term $(p_1+m)/2m$ in (29) is effectively 1 and we are left with $-r$ of (23).[16]

In more complex problems starting with a free electron the same type of term arises from the effects of a virtual emission and absorption both previous to the other processes. They, therefore, simply lead to the same factor r so that the expression (23) may be used directly and these renormalization integrals need not be computed afresh for each problem.

In this problem of the radiative corrections to scattering the net result is insensitive to the cut-off. This means, of course, that by a simple rearrangement of terms previous to the integration we could have avoided the use of the convergence factors completely (see for example Lewis[17]). The problem was solved in the manner here in order to illustrate how the use of such convergence factors, even when they are actually unnecessary, may facilitate analysis somewhat by removing the effort and ambiguities that may be involved in trying to rearrange the otherwise divergent terms.

The replacement of δ_+ by f_+ given in (16), (17) is not determined by the analogy with the classical problem. In the classical limit only the real part of δ_+ (i.e., just δ) is easy to interpret. But by what should the imaginary part, $1/(\pi i s^2)$, of δ_+ be replaced? The choice we have made here (in defining, as we have, the location of the poles of (17)) is arbitrary and almost certainly incorrect. If the radiation resistance is calculated for an atom, as the imaginary part of (8), the result depends slightly on the function f_+. On the other hand the light radiated at very large distances from a source is independent of f_+. The total energy absorbed by distant absorbers will not check with the energy loss of the source. We are in a situation analogous to that in the classical theory if the entire f function is made to contain only retarded contributions (see A, Appendix). One desires instead the analogue of $\langle F\rangle_{\text{ret}}$ of A. This problem is being studied.

One can say therefore, that this attempt to find a consistent modification of quantum electrodynamics is incomplete (see also the question of closed loops, below). For it could turn out that any correct form of f_+ which will guarantee energy conservation may at the same time not be able to make the self-energy integral finite. The desire to make the methods of simplifying the calculation of quantum electrodynamic processes more widely available has prompted this publication before an analysis of the correct form for f_+ is complete. One might try to take the position that, since the energy discrepancies discussed vanish in the limit $\lambda\to\infty$, the correct physics might be considered to be that obtained by letting $\lambda\to\infty$ after mass renormalization. I have no proof of the mathematical consistency of this procedure, but the presumption is very strong that it is satisfactory. (It is also strong that a satisfactory form for f_+ can be found.)

7. THE PROBLEM OF VACUUM POLARIZATION

In the analysis of the radiative corrections to scattering one type of term was not considered. The potential

[14] The expression is not exact because the substitution of Δm by the integral in (19) is valid only if p operates on a state such that p can be replaced by m. The error, however, is of order $a(p'-m)^{-1}(p-m)(p'-m)^{-1}b$ which is $a((1+\epsilon)p+m)(p-m)\times((1+\epsilon)p+m)p(2\epsilon+\epsilon^2)^{-2}m^{-4}$. But since $p^2=m^2$, we have $p(p-m)=-m(p-m)=(p-m)p$ so the net result is approximately $a(p-m)b/4m^2$ and is not of order $1/\epsilon$ but smaller, so that its effect drops out in the limit.

[15] We have used, to first order, the general expansion (valid for any operators A, B)

$$(A+B)^{-1}=A^{-1}-A^{-1}BA^{-1}+A^{-1}BA^{-1}BA^{-1}-\cdots$$

with $A=p-k-m$ and $B=p'-p=\epsilon p$ to expand the difference of $(p'-k-m)^{-1}$ and $(p-k-m)^{-1}$.

[16] The renormalization terms appearing B, Eqs. (14), (15) when translated directly into the present notation do not give twice (29) but give this expression with the central p_1m^{-1} factor replaced by $m\gamma_4/E_1$ where $E_1=p_{1\mu}$ for $\mu=4$. When integrated it therefore gives $ra((p_1+m)/2m)(m\gamma_4/E_1)$ or $ra-ra(m\gamma_4/E_1)(p_1-m)/2m$. (Since $p_1\gamma_4+\gamma_4p_1=2E_1$) which gives just ra, since $p_1u_1=mu_1$.

[17] H. W. Lewis, Phys. Rev. 73, 173 (1948).

which we can assume to vary as $a_\mu \exp(-iq\cdot x)$ creates a pair of electrons (see Fig. 6), momenta p_a, $-p_b$. This pair then reannihilates, emitting a quantum $q=p_b-p_a$, which quantum scatters the original electron from state 1 to state 2. The matrix element for this process (and the others which can be obtained by rearranging the order in time of the various events) is

$$-(e^2/\pi i)(\bar{u}_2\gamma_\mu u_1)\int Sp[(p_a+q-m)^{-1} \times\gamma_\nu(p_a-m)^{-1}\gamma_\mu]d^4p_a q^{-2}C(q^2)a_\nu. \quad (30)$$

This is because the potential produces the pair with amplitude proportional to $a_\nu\gamma_\nu$, the electrons of momenta p_a and $-(p_a+q)$ proceed from there to annihilate, producing a quantum (factor γ_μ) which propagates (factor $q^{-2}C(q^2)$) over to the other electron, by which it is absorbed (matrix element of γ_μ between states 1 and 2 of the original electron $(\bar{u}_2\gamma_\mu u_1)$). All momenta p_a and spin states of the virtual electron are admitted, which means the spur and the integral on d^4p_a are calculated.

One can imagine that the closed loop path of the positron-electron produces a current

$$4\pi j_\mu = J_{\mu\nu}a_\nu, \quad (31)$$

which is the source of the quanta which act on the second electron. The quantity

$$J_{\mu\nu}=-(e^2/\pi i)\int Sp[(p+q-m)^{-1} \times\gamma_\nu(p-m)^{-1}\gamma_\mu]d^4p, \quad (32)$$

is then characteristic for this problem of polarization of the vacuum.

One sees at once that $J_{\mu\nu}$ diverges badly. The modification of δ to f alters the amplitude with which the current j_μ will affect the scattered electron, but it can do nothing to prevent the divergence of the integral (32) and of its effects.

One way to avoid such difficulties is apparent. From one point of view we are considering all routes by which a given electron can get from one region of space-time to another, i.e., from the source of electrons to the apparatus which measures them. From this point of view the closed loop path leading to (32) is unnatural. It might be assumed that the only paths of meaning are those which start from the source and work their way in a continuous path (possibly containing many time reversals) to the detector. Closed loops would be excluded. We have already found that this may be done for electrons moving in a fixed potential.

Such a suggestion must meet several questions, however. The closed loops are a consequence of the usual hole theory in electrodynamics. Among other things, they are required to keep probability conserved. The probability that no pair is produced by a potential is

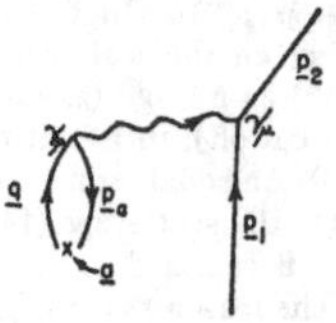

FIG. 6. Vacuum polarization effect on scattering, Eq. (30).

not unity and its deviation from unity arises from the imaginary part of $J_{\mu\nu}$. Again, with closed loops excluded, a pair of electrons once created cannot annihilate one another again, the scattering of light by light would be zero, etc. Although we are not experimentally sure of these phenomena, this does seem to indicate that the closed loops are necessary. To be sure, it is always possible that these matters of probability conservation, etc., will work themselves out as simply in the case of interacting particles as for those in a fixed potential. Lacking such a demonstration the presumption is that the difficulties of vacuum polarization are not so easily circumvented.[18]

An alternative procedure discussed in *B* is to assume that the function $K_+(2,1)$ used above is incorrect and is to be replaced by a modified function K_+' having no singularity on the light cone. The effect of this is to provide a convergence factor $C(p^2-m^2)$ for *every* integral over electron momenta.[19] This will multiply the integrand of (32) by $C(p^2-m^2)C((p+q)^2-m^2)$, since the integral was originally $\delta(p_a-p_b+q)d^4p_a d^4p_b$ and both p_a and p_b get convergence factors. The integral now converges but the result is unsatisfactory.[20]

One expects the current (31) to be conserved, that is $q_\mu j_\mu=0$ or $q_\mu J_{\mu\nu}=0$. Also one expects no current if a_ν is a gradient, or $a_\nu=q_\nu$ times a constant. This leads to the condition $J_{\mu\nu}q_\nu=0$ which is equivalent to $q_\mu J_{\mu\nu}=0$ since $J_{\mu\nu}$ is symmetrical. But when the expression (32) is integrated with such convergence factors it does not satisfy this condition. By altering the kernel from K to another, K', which does not satisfy the Dirac equation we have lost the gauge invariance, its consequent current conservation and the general consistency of the theory.

One can see this best by calculating $J_{\mu\nu}q_\nu$ directly from (32). The expression within the spur becomes $(p+q-m)^{-1}q(p-m)^{-1}\gamma_\mu$ which can be written as the difference of two terms: $(p-m)^{-1}\gamma_\mu-(p+q-m)^{-1}\gamma_\mu$. Each of these terms would give the same result if the integration d^4p were without a convergence factor, for

[18] It would be very interesting to calculate the Lamb shift accurately enough to be sure that the 20 megacycles expected from vacuum polarization are actually present.

[19] This technique also makes self-energy and radiationless scattering integrals finite even without the modification of δ_+ to f_+ for the radiation (and the consequent convergence factor $C(k^2)$ for the quanta). See B.

[20] Added to the terms given below (33) there is a term $\frac{1}{2}(\lambda^2-2\mu^2+\frac{1}{3}q^2)\delta_{\mu\nu}$ for $C(k^2)=-\lambda^2(k^2-\lambda^2)^{-1}$, which is not gauge invariant. (In addition the charge renormalization has $-7/6$ added to the logarithm.)

the first can be converted into the second by a shift of the origin of p, namely $p'=p+q$. This does not result in cancelation in (32) however, for the convergence factor is altered by the substitution.

A method of making (32) convergent without spoiling the gauge invariance has been found by Bethe and by Pauli. The convergence factor for light can be looked upon as the result of superposition of the effects of quanta of various masses (some contributing negatively). Likewise if we take the factor $C(p^2-m^2) = -\lambda^2(p^2-m^2-\lambda^2)^{-1}$ so that $(p^2-m^2)^{-1}C(p^2-m^2) = (p^2-m^2)^{-1}-(p^2-m^2-\lambda^2)^{-1}$ we are taking the difference of the result for electrons of mass m and mass $(\lambda^2+m^2)^{\frac{1}{2}}$. But we have taken this difference for *each* propagation between interactions with photons. They suggest instead that once created with a certain mass the electron should continue to propagate with this mass through all the potential interactions until it closes its loop. That is if the quantity (32), integrated over some finite range of p, is called $J_{\mu\nu}(m^2)$ and the corresponding quantity over the same range of p, but with m replaced by $(m^2+\lambda^2)^{\frac{1}{2}}$ is $J_{\mu\nu}(m^2+\lambda^2)$ we should calculate

$$J_{\mu\nu}{}^P=\int_0^\infty [J_{\mu\nu}(m^2)-J_{\mu\nu}(m^2+\lambda^2)]G(\lambda)d\lambda, \quad (32')$$

the function $G(\lambda)$ satisfying $\int_0^\infty G(\lambda)d\lambda=1$ and $\int_0^\infty G(\lambda)\lambda^2 d\lambda=0$. Then in the expression for $J_{\mu\nu}{}^P$ the range of p integration can be extended to infinity as the integral now converges. The result of the integration using this method is the integral on $d\lambda$ over $G(\lambda)$ of (see Appendix C)

$$J_{\mu\nu}{}^P=-\frac{e^2}{\pi}(q_\mu q_\nu-\delta_{\mu\nu}q^2)\bigg(-\frac{1}{3}\ln\frac{\lambda^2}{m^2} -\bigg[\frac{4m^2+2q^2}{3q^2}\bigg(1-\frac{\theta}{\tan\theta}\bigg)-\frac{1}{9}\bigg]\bigg), \quad (33)$$

with $q^2=4m^2\sin^2\theta$.

The gauge invariance is clear, since $q_\mu(q_\mu q_\nu-q^2\delta_{\mu\nu})=0$. Operating (as it always will) on a potential of zero divergence the $(q_\mu q_\nu-\delta_{\mu\nu}q^2)a_\nu$ is simply $-q^2 a_\mu$, the D'Alembertian of the potential, that is, the current producing the potential. The term $-\frac{1}{3}(\ln(\lambda^2/m^2))(q_\mu q_\nu -q^2\delta_{\mu\nu})$ therefore gives a current proportional to the current producing the potential. This would have the same effect as a change in charge, so that we would have a difference $\Delta(e^2)$ between e^2 and the experimentally observed charge, $e^2+\Delta(e^2)$, analogous to the difference between m and the observed mass. This charge depends logarithmically on the cut-off, $\Delta(e^2)/e^2= -(2e^2/3\pi)\ln(\lambda/m)$. After this renormalization of charge is made, no effects will be sensitive to the cut-off.

After this is done the final term remaining in (33), contains the usual effects[21] of polarization of the vacuum. It is zero for a free light quantum ($q^2=0$). For small q^2 it behaves as $(2/15)q^2$ (adding $-\frac{1}{5}$ to the logarithm in the Lamb effect). For $q^2>(2m)^2$ it is complex, the imaginary part representing the loss in amplitude required by the fact that the probability that no quanta are produced by a potential able to produce pairs $((q^2)^{\frac{1}{2}}>2m)$ decreases with time. (To make the necessary analytic continuation, imagine m to have a small negative imaginary part, so that $(1-q^2/4m^2)^{\frac{1}{2}}$ becomes $-i(q^2/4m^2-1)^{\frac{1}{2}}$ as q^2 goes from below to above $4m^2$. Then $\theta=\pi/2+iu$ where $\sinh u=+(q^2/4m^2-1)^{\frac{1}{2}}$, and $-1/\tan\theta=i\tanh u=+i(q^2-4m^2)^{\frac{1}{2}}(q^2)^{-\frac{1}{2}}$.)

Closed loops containing a number of quanta or potential interactions larger than two produce no trouble. Any loop with an odd number of interactions gives zero (I, reference 9). Four or more potential interactions give integrals which are convergent even without a convergence factor as is well known. The situation is analogous to that for self-energy. Once the simple problem of a single closed loop is solved there are no further divergence difficulties for more complex processes.[22]

8. LONGITUDINAL WAVES

In the usual form of quantum electrodynamics the longitudinal and transverse waves are given separate treatment. Alternately the condition $(\partial A_\mu/\partial x_\mu)\Psi=0$ is carried along as a supplementary condition. In the present form no such special considerations are necessary for we are dealing with the solutions of the equation $-\Box^2 A_\mu=4\pi j_\mu$ with a current j_μ which is conserved $\partial j_\mu/\partial x_\mu=0$. That means at least $\Box^2(\partial A_\mu/\partial x_\mu)=0$ and in fact our solution also satisfies $\partial A_\mu/\partial x_\mu=0$.

To show that this is the case we consider the amplitude for emission (real or virtual) of a photon and show that the divergence of this amplitude vanishes. The amplitude for emission for photons polarized in the μ direction involves matrix elements of γ_μ. Therefore what we have to show is that the corresponding matrix elements of $q_\mu\gamma_\mu=q$ vanish. For example, for a first order effect we would require the matrix element of q between two states p_1 and $p_2=p_1+q$. But since $q=p_2-p_1$ and $(\bar{u}_2 p_1 u_1)=m(\bar{u}_2 u_1)=(\bar{u}_2 p_2 u_1)$ the matrix element vanishes, which proves the contention in this case. It also vanishes in more complex situations (essentially because of relation (34), below) (for example, try putting $e_2=q_2$ in the matrix (15) for the Compton Effect).

To prove this in general, suppose a_i, $i=1$ to N are a set of plane wave disturbing potentials carrying momenta q_i (e.g., some may be emissions or absorptions of the same or different quanta) and consider a matrix for the transition from a state of momentum p_0 to p_N such

[21] E. A. Uehling, Phys. Rev. 48, 55 (1935), R. Serber, Phys. Rev. 48, 49 (1935).

[22] There are loops completely without external interactions. For example, a pair is created virtually along with a photon. Next they annihilate, absorbing this photon. Such loops are disregarded on the grounds that they do not interact with anything and are thereby completely unobservable. Any indirect effects they may have via the exclusion principle have already been included.

as $a_N \prod_{i=1}^{N-1} (p_i - m)^{-1} a_i$ where $p_i = p_{i-1} + q_i$ (and in the product, terms with larger i are written to the left). The most general matrix element is simply a linear combination of these. Next consider the matrix between states p_0 and $p_N + q$ in a situation in which not only are the a_i acting but also another potential $a \exp(-iq\cdot x)$ where $a = q$. This may act previous to all a_i, in which case it gives $a_N \prod (p_i + q - m)^{-1} a_i (p_0 + q - m)^{-1} q$ which is equivalent to $+a_N \prod (p_i + q - m)^{-1} a_i$ since $+(p_0 + q - m)^{-1} q$ is equivalent to $(p_0 + q - m)^{-1} \times (p_0 + q - m)$ as p_0 is equivalent to m acting on the initial state. Likewise if it acts after all the potentials it gives $q(p_N - m)^{-1} a_N \prod (p_i - m)^{-1} a_i$ which is equivalent to $-a_N \prod (p_i - m)^{-1} a_i$ since $p_N + q - m$ gives zero on the final state. Or again it may act between the potential a_k and a_{k+1} for each k. This gives

$$\sum_{k=1}^{N-1} a_N \prod_{i=k+1}^{N-1} (p_i + q - m)^{-1} a_i (p_k + q - m)^{-1} \times q (p_k - m)^{-1} a_k \prod_{j=1}^{k-1} (p_j - m)^{-1} a_j.$$

However,

$$(p_k + q - m)^{-1} q (p_k - m)^{-1} = (p_k - m)^{-1} - (p_k + q - m)^{-1}, \quad (34)$$

so that the sum breaks into the difference of two sums, the first of which may be converted to the other by the replacement of k by $k-1$. There remain only the terms from the ends of the range of summation,

$$+a_N \prod_{i=1}^{N-1} (p_i - m)^{-1} a_i - a_N \prod_{i=1}^{N-1} (p_i + q - m)^{-1} a_i.$$

These cancel the two terms originally discussed so that the entire effect is zero. Hence any wave emitted will satisfy $\partial A_\mu / \partial x_\mu = 0$. Likewise longitudinal waves (that is, waves for which $A_\mu = \partial \phi / \partial x_\mu$ or $a = q$) cannot be absorbed and will have no effect, for the matrix elements for emission and absorption are similar. (We have said little more than that a potential $A_\mu = \partial \varphi / \partial x_\mu$ has no effect on a Dirac electron since a transformation $\psi' = \exp(-i\phi)\psi$ removes it. It is also easy to see in coordinate representation using integrations by parts.)

This has a useful practical consequence in that in computing probabilities for transition for unpolarized light one can sum the squared matrix over all four directions rather than just the two special polarization vectors. Thus suppose the matrix element for some process for light polarized in direction e_μ is $e_\mu M_\mu$. If the light has wave vector q_μ we know from the argument above that $q_\mu M_\mu = 0$. For unpolarized light progressing in the z direction we would ordinarily calculate $M_x^2 + M_y^2$. But we can as well sum $M_x^2 + M_y^2 + M_z^2 - M_t^2$ for $q_\mu M_\mu$ implies $M_t = M_z$ since $q_t = q_z$ for free quanta. This shows that unpolarized light is a relativistically invariant concept, and permits some simplification in computing cross sections for such light.

Incidentally, the virtual quanta interact through terms like $\gamma_\mu \cdots \gamma_\mu k^{-2} d^4k$. Real processes correspond to poles in the formulae for virtual processes. The pole occurs when $k^2 = 0$, but it looks at first as though in the sum on all four values of μ, of $\gamma_\mu \cdots \gamma_\mu$ we would have four kinds of polarization instead of two. Now it is clear that only two perpendicular to k are effective.

The usual elimination of longitudinal and scalar virtual photons (leading to an instantaneous Coulomb potential) can of course be performed here too (although it is not particularly useful). A typical term in a virtual transition is $\gamma_\mu \cdots \gamma_\mu k^{-2} d^4k$ where the $\cdots$ represent some intervening matrices. Let us choose for the values of μ, the time t, the direction of vector part $\mathbf{K}$, of k, and two perpendicular directions 1, 2. We shall not change the expression for these two 1, 2 for these are represented by transverse quanta. But we must find $(\gamma_t \cdots \gamma_t) - (\gamma_{\mathbf{K}} \cdots \gamma_{\mathbf{K}})$. Now $k = k_4 \gamma_t - K \gamma_{\mathbf{K}}$, where $K = (\mathbf{K}\cdot\mathbf{K})^{\frac{1}{2}}$, and we have shown above that k replacing the γ_μ gives zero.[23] Hence $K\gamma_{\mathbf{K}}$ is equivalent to $k_4 \gamma_t$ and

$$(\gamma_t \cdots \gamma_t) - (\gamma_{\mathbf{K}} \cdots \gamma_{\mathbf{K}}) = ((K^2 - k_4^2)/K^2)(\gamma_t \cdots \gamma_t),$$

so that on multiplying by $k^{-2} d^4k = d^4k (k_4^2 - K^2)^{-1}$ the net effect is $-(\gamma_t \cdots \gamma_t) d^4k / K^2$. The γ_t means just scalar waves, that is, potentials produced by charge density. The fact that $1/K^2$ does not contain k_4 means that k_4 can be integrated first, resulting in an instantaneous interaction, and the $d^3\mathbf{K}/K^2$ is just the momentum representation of the Coulomb potential, $1/r$.

9. KLEIN GORDON EQUATION

The methods may be readily extended to particles of spin zero satisfying the Klein Gordon equation,[24]

$$\Box^2 \psi - m^2 \psi = i\partial(A_\mu \psi)/\partial x_\mu + iA_\mu \partial\psi/\partial x_\mu - A_\mu A_\mu \psi. \quad (35)$$

---

[23] A little more care is required when both γ_μ's act on the same particle. Define $x = k_4 \gamma_t + K \gamma_{\mathbf{K}}$, and consider $(k \cdots x) + (x \cdots k)$. Exactly this term would arise if a system, acted on by potential x carrying momentum $-k$, is disturbed by an added potential k of momentum $+k$ (the reversed sign of the momenta in the intermediate factors in the second term $x \cdots k$ has no effect since we will later integrate over all k). Hence as shown above the result is zero, but since $(k \cdots x) + (x \cdots k) = k_4^2 (\gamma_t \cdots \gamma_t) - K^2 (\gamma_{\mathbf{K}} \cdots \gamma_{\mathbf{K}})$ we can still conclude $(\gamma_{\mathbf{K}} \cdots \gamma_{\mathbf{K}}) = k_4^2 K^{-2} (\gamma_t \cdots \gamma_t)$.

[24] The equations discussed in this section were deduced from the formulation of the Klein Gordon equation given in reference 5, Section 14. The function ψ in this section has only one component and is not a spinor. An alternative formal method of making the equations valid for spin zero and also for spin 1 is (presumably) by use of the Kemmer-Duffin matrices β_μ, satisfying the commutation relation

$$\beta_\mu \beta_\nu \beta_\sigma + \beta_\sigma \beta_\nu \beta_\mu = \delta_{\mu\nu} \beta_\sigma + \delta_{\sigma\nu} \beta_\mu.$$

If we interpret a to mean $a_\mu \beta_\mu$, rather than $a_\mu \gamma_\mu$, for any a_μ, all of the equations in momentum space will remain formally identical to those for the spin 1/2; with the exception of those in which a denominator $(p-m)^{-1}$ has been rationalized to $(p+m)(p^2-m^2)^{-1}$ since p^2 is no longer equal to a number, $p\cdot p$. But p^3 does equal $(p\cdot p)p$ so that $(p-m)^{-1}$ may now be interpreted as $(mp + m^2 + p^2 - p\cdot p)(p\cdot p - m^2)^{-1} m^{-1}$. This implies that equations in coordinate space will be valid of the function $K_+(2,1)$ is given as $K_+(2,1) = [(i\nabla_2 + m) - m^{-1}(\nabla_2^2 + \Box_2^2)] iI_+(2,1)$ with $\nabla_2 = \beta_\mu \partial/\partial x_{2\mu}$. This is all in virtue of the fact that the many component wave function ψ (5 components for spin 0, 10 for spin 1) satisfies $(i\nabla - m)\psi = A\psi$ which is formally identical to the Dirac Equation. See W. Pauli, Rev. Mod. Phys. 13, 203 (1940).

The important kernel is now $I_+(2,1)$ defined in (I, Eq. (32)). For a free particle, the wave function $\psi(2)$ satisfies $+\Box^2\psi-m^2\psi=0$. At a point, 2, inside a space time region it is given by

$$\psi(2)=\int[\psi(1)\partial I_+(2,1)/\partial x_{1\mu}\\ -(\partial\psi/\partial x_{1\mu})I_+(2,1)]N_\mu(1)d^3V_1,$$

(as is readily shown by the usual method of demonstrating Green's theorem) the integral being over an entire 3-surface boundary of the region (with normal vector N_μ). Only the positive frequency components of ψ contribute from the surface preceding the time corresponding to 2, and only negative frequencies from the surface future to 2. These can be interpreted as electrons and positrons in direct analogy to the Dirac case.

The right-hand side of (35) can be considered as a source of new waves and a series of terms written down to represent matrix elements for processes of increasing order. There is only one new point here, the term in $A_\mu A_\mu$ by which two quanta can act at the same time. As an example, suppose three quanta or potentials, $a_\mu\exp(-iq_a\cdot x)$, $b_\mu\exp(-iq_b\cdot x)$, and $c_\mu\exp(-iq_c\cdot x)$ are to act in that order on a particle of original momentum $p_{0\mu}$ so that $p_a=p_0+q_a$ and $p_b=p_a+q_b$; the final momentum being $p_c=p_b+q_c$. The matrix element is the sum of three terms ($p^2=p_\mu p_\mu$) (illustrated in Fig. 7)

$$\begin{aligned}&(p_c\cdot c+p_b\cdot c)(p_b^2-m^2)^{-1}(p_b\cdot b+p_a\cdot b)\\ &\qquad\times(p_a^2-m^2)^{-1}(p_a\cdot a+p_0\cdot a)\\ &-(p_c\cdot c+p_b\cdot c)(p_b^2-m^2)^{-1}(b\cdot a)\\ &-(c\cdot b)(p_a^2-m^2)^{-1}(p_a\cdot a+p_0\cdot a).\end{aligned}\tag{36}$$

The first comes when each potential acts through the perturbation $i\partial(A_\mu\psi)/\partial x_\mu+iA_\mu\partial\psi/\partial x_\mu$. These gradient operators in momentum space mean respectively the momentum after and before the potential A_μ operates. The second term comes from b_μ and a_μ acting at the same instant and arises from the $A_\mu A_\mu$ term in (a). Together b_μ and a_μ carry momentum $q_{b\mu}+q_{a\mu}$ so that after $b\cdot a$ operates the momentum is $p_0+q_a+q_b$ or p_b. The final term comes from c_μ and b_μ operating together in a similar manner. The term $A_\mu A_\mu$ thus permits a new type of process in which two quanta can be emitted (or absorbed, or one absorbed, one emitted) at the same time. There is no $a\cdot c$ term for the order a, b, c we have assumed. In an actual problem there would be other terms like (36) but with alterations in the order in which the quanta a, b, c act. In these terms $a\cdot c$ would appear.

As a further example the self-energy of a particle of momentum p_μ is

$$(e^2/2\pi im)\int[(2p-k)_\mu((p-k)^2-m^2)^{-1}\\ \times(2p-k)_\mu-\delta_{\mu\mu}]d^4kk^{-2}C(k^2),$$

where the $\delta_{\mu\mu}=4$ comes from the $A_\mu A_\mu$ term and represents the possibility of the simultaneous emission and absorption of the same virtual quantum. This integral without the $C(k^2)$ diverges quadratically and would not converge if $C(k^2)=-\lambda^2/(k^2-\lambda^2)$. Since the interaction occurs through the gradients of the potential, we must use a stronger convergence factor, for example $C(k^2)=\lambda^4(k^2-\lambda^2)^{-2}$, or in general (17) with $\int_0^\infty\lambda^2G(\lambda)d\lambda=0$. In this case the self-energy converges but depends quadratically on the cut-off λ and is not necessarily small compared to m. The radiative corrections to scattering after mass renormalization are insensitive to the cut-off just as for the Dirac equation.

When there are several particles one can obtain Bose statistics by the rule that if two processes lead to the same state but with two electrons exchanged, their amplitudes are to be added (rather than subtracted as for Fermi statistics). In this case equivalence to the second quantization treatment of Pauli and Weisskopf should be demonstrable in a way very much like that given in I (appendix) for Dirac electrons. The Bose statistics mean that the sign of contribution of a closed loop to the vacuum polarization is the opposite of what it is for the Fermi case (see I). It is ($p_b=p_a+q$)

$$J_{\mu\nu}=\frac{e^2}{2\pi im}\int[(p_{b\mu}+p_{a\mu})(p_{b\nu}+p_{a\nu})(p_a^2-m^2)^{-1}\\ \times(p_b^2-m^2)^{-1}-\delta_{\mu\nu}(p_a^2-m^2)^{-1}\\ -\delta_{\mu\nu}(p_b^2-m^2)^{-1}]d^4p_a$$

giving,

$$J_{\mu\nu}{}^P=\frac{e^2}{\pi}(q_\mu q_\nu-\delta_{\mu\nu}q^2)\left[\frac{1}{6}\ln\frac{\lambda^2}{m^2}+\frac{1}{9}-\frac{4m^2-q^2}{3q^2}\left(1-\frac{\theta}{\tan\theta}\right)\right],$$

the notation as in (33). The imaginary part for $(q^2)^{\frac{1}{2}}>2m$ is again positive representing the loss in the probability of finding the final state to be a vacuum, associated with the possibilities of pair production. Fermi statistics would give a gain in probability (and also a charge renormalization of opposite sign to that expected).

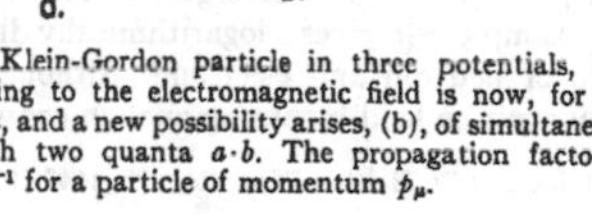

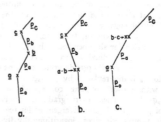

FIG. 7. Klein-Gordon particle in three potentials, Eq. (36). The coupling to the electromagnetic field is now, for example, $p_0\cdot a+p_a\cdot a$, and a new possibility arises, (b), of simultaneous interaction with two quanta $a\cdot b$. The propagation factor is now $(p\cdot p-m^2)^{-1}$ for a particle of momentum p_μ.

10. APPLICATION TO MESON THEORIES

The theories which have been developed to describe mesons and the interaction of nucleons can be easily expressed in the language used here. Calculations, to lowest order in the interactions can be made very easily for the various theories, but agreement with experimental results is not obtained. Most likely all of our present formulations are quantitatively unsatisfactory. We shall content ourselves therefore with a brief summary of the methods which can be used.

The nucleons are usually assumed to satisfy Dirac's equation so that the factor for propagation of a nucleon of momentum p is $(p-M)^{-1}$ where M is the mass of the nucleon (which implies that nucleons can be created in pairs). The nucleon is then assumed to interact with mesons, the various theories differing in the form assumed for this interaction.

First, we consider the case of neutral mesons. The theory closest to electrodynamics is the theory of vector mesons with vector coupling. Here the factor for emission or absorption of a meson is $g\gamma_\mu$ when this meson is "polarized" in the μ direction. The factor g, the "mesonic charge," replaces the electric charge e. The amplitude for propagation of a meson of momentum q in intermediate states is $(q^2-\mu^2)^{-1}$ (rather than q^{-2} as it is for light) where μ is the mass of the meson. The necessary integrals are made finite by convergence factors $C(q^2-\mu^2)$ as in electrodynamics. For scalar mesons with scalar coupling the only change is that one replaces the γ_μ by 1 in emission and absorption. There is no longer a direction of polarization, μ, to sum upon. For pseudoscalar mesons, pseudoscalar coupling replace γ_μ by $\gamma_5=i\gamma_x\gamma_y\gamma_z\gamma_t$. For example, the self-energy matrix of a nucleon of momentum p in this theory is

$$(g^2/\pi i)\int \gamma_5(p-k-M)^{-1}\gamma_5 d^4k(k^2-\mu^2)^{-1}C(k^2-\mu^2).$$

Other types of meson theory result from the replacement of γ_μ by other expressions (for example by $\frac{1}{2}(\gamma_\mu\gamma_\nu-\gamma_\nu\gamma_\mu)$ with a subsequent sum over all μ and ν for virtual mesons). Scalar mesons with vector coupling result from the replacement of γ_μ by $\mu^{-1}q$ where q is the final momentum of the nucleon minus its initial momentum, that is, it is the momentum of the meson if absorbed, or the negative of the momentum of a meson emitted. As is well known, this theory with neutral mesons gives zero for all processes, as is proved by our discussion on longitudinal waves in electrodynamics. Pseudoscalar mesons with pseudo-vector coupling corresponds to γ_μ being replaced by $\mu^{-1}\gamma_5 q$ while vector mesons with tensor coupling correspond to using $(2\mu)^{-1}(\gamma_\mu q-q\gamma_\mu)$. These extra gradients involve the danger of producing higher divergencies for real processes. For example, $\gamma_5 q$ gives a logarithmically divergent interaction of neutron and electron.[25] Although these divergencies can be held by strong enough convergence factors, the results then are sensitive to the method used for convergence and the size of the cut-off values of λ. For low order processes $\mu^{-1}\gamma_5 q$ is equivalent to the pseudoscalar interaction $2M\mu^{-1}\gamma_5$ because if taken between free particle wave functions of the nucleon of momenta p_1 and $p_2=p_1+q$, we have

$$(\bar{u}_2\gamma_5 q u_1)=(\bar{u}_2\gamma_5(p_2-p_1)u_1)=-(\bar{u}_2 p_2\gamma_5 u_1)-(\bar{u}_2\gamma_5 p_1 u_1)=-2M(\bar{u}_2\gamma_5 u_1)$$

since γ_5 anticommutes with p_2 and p_2 operating on the state 2 equivalent to M as is p_1 on the state 1. This shows that the γ_5 interaction is unusually weak in the non-relativistic limit (for example the expected value of γ_5 for a free nucleon is zero), but since $\gamma_5^2=1$ is not small, pseudoscalar theory gives a more important interaction in second order than it does in first. Thus the pseudoscalar coupling constant should be chosen to fit nuclear forces including these important second order processes.[26] The equivalence of pseudoscalar and pseudovector coupling which holds for low order processes therefore does not hold when the pseudoscalar theory is giving its most important effects. These theories will therefore give quite different results in the majority of practical problems.

In calculating the corrections to scattering of a nucleon by a neutral vector meson field (γ_μ) due to the effects of virtual mesons, the situation is just as in electrodynamics, in that the result converges without need for a cut-off and depends only on gradients of the meson potential. With scalar (1) or pseudoscalar (γ_5) neutral mesons the result diverges logarithmically and so must be cut off. The part sensitive to the cut-off, however, is directly proportional to the meson potential. It may thereby be removed by a renormalization of mesonic charge g. After this renormalization the results depend only on gradients of the meson potential and are essentially independent of cut-off. This is in addition to the mesonic charge renormalization coming from the production of virtual nucleon pairs by a meson, analogous to the vacuum polarization in electrodynamics. But here there is a further difference from electrodynamics for scalar or pseudoscalar mesons in that the polarization also gives a term in the induced current proportional to the meson potential representing therefore an additional renormalization of the *mass of the meson* which usually depends quadratically on the cut-off.

Next consider charged mesons in the absence of an electromagnetic field. One can introduce isotopic spin operators in an obvious way. (Specifically replace the neutral γ_5, say, by $\tau_i\gamma_5$ and sum over $i=1, 2$ where $\tau_1=\tau_++\tau_-$, $\tau_2=i(\tau_+-\tau_-)$ and τ_+ changes neutron to proton (τ_+ on proton$=0$) and τ_- changes proton to neutron.) It is just as easy for practical problems simply to keep track of whether the particle is a proton or a neutron on a diagram drawn to help write down the

[25] M. Slotnick and W. Heitler, Phys. Rev. 75, 1645 (1949).

[26] H. A. Bethe, Bull. Am. Phys. Soc. 24, 3, Z3 (Washington, 1949).

matrix element. This excludes certain processes. For example in the scattering of a negative meson from q_1 to q_2 by a neutron, the meson q_2 must be emitted first (in order of operators, not time) for the neutron cannot absorb the negative meson q_1 until it becomes a proton. That is, in comparison to the Klein Nishina formula (15), only the analogue of second term (see Fig. 5(b)) would appear in the scattering of negative mesons by neutrons, and only the first term (Fig. 5(a)) in the neutron scattering of positive mesons.

The source of mesons of a given charge is not conserved, for a neutron capable of emitting negative mesons may (on emitting one, say) become a proton no longer able to do so. The proof that a perturbation q gives zero, discussed for longitudinal electromagnetic waves, fails. This has the consequence that vector mesons, if represented by the interaction γ_μ would not satisfy the condition that the divergence of the potential is zero. The interaction is to be taken[27] as $\gamma_\mu - \mu^{-2}q_\mu q$ in emission and as γ_μ in absorption if the real emission of mesons with a non-zero divergence of potential is to be avoided. (The correction term $\mu^{-2}q_\mu q$ gives zero in the neutral case.) The asymmetry in emission and absorption is only apparent, as this is clearly the same thing as subtracting from the original $\gamma_\mu \cdots \gamma_\mu$, a term $\mu^{-2}q \cdots q$. That is, if the term $-\mu^{-2}q_\mu q$ is omitted the resulting theory describes a combination of mesons of spin one and spin zero. The spin zero mesons, coupled by vector coupling q, are removed by subtracting the term $\mu^{-2}q \cdots q$.

The two extra gradients $q \cdots q$ make the problem of diverging integrals still more serious (for example the interaction between two protons corresponding to the exchange of two charged vector mesons depends quadratically on the cut-off if calculated in a straightforward way). One is tempted in this formulation to choose simply $\gamma_\mu \cdots \gamma_\mu$ and accept the admixture of spin zero mesons. But it appears that this leads in the conventional formalism to negative energies for the spin zero component. This shows one of the advantages of the method of second quantization of meson fields over the present formulation. There such errors of sign are obvious while here we seem to be able to write seemingly innocent expressions which can give absurd results. Pseudovector mesons with pseudovector coupling correspond to using $\gamma_5(\gamma_\mu - \mu^{-2}q_\mu q)$ for absorption and $\gamma_5\gamma_\mu$ for emission for both charged and neutral mesons.

In the presence of an electromagnetic field, whenever the nucleon is a proton it interacts with the field in the way described for electrons. The meson interacts in the scalar or pseudoscalar case as a particle obeying the Klein-Gordon equation. It is important here to use the method of calculation of Bethe and Pauli, that is, a virtual meson is assumed to have the same "mass" during all its interactions with the electromagnetic field. The result for mass μ and for $(\mu^2+\lambda^2)^{\frac{1}{2}}$ are subtracted and the difference integrated over the function $G(\lambda)d\lambda$. A separate convergence factor is not provided for each meson propagation between electromagnetic interactions, otherwise gauge invariance is not insured. When the coupling involves a gradient, such as $\gamma_5 q$ where q is the final minus the initial momentum of the nucleon, the vector potential A must be subtracted from the momentum of the proton. That is, there is an additional coupling $\pm\gamma_5 A$ (plus when going from proton to neutron, minus for the reverse) representing the new possibility of a simultaneous emission (or absorption) of meson and photon.

Emission of positive or absorption of negative virtual mesons are represented in the same term, the sign of the charge being determined by temporal relations as for electrons and positrons.

Calculations are very easily carried out in this way to lowest order in g^2 for the various theories for nucleon interaction, scattering of mesons by nucleons, meson production by nuclear collisions and by gamma-rays, nuclear magnetic moments, neutron electron scattering, etc., However, no good agreement with experiment results, when these are available, is obtained. Probably all of the formulations are incorrect. An uncertainty arises since the calculations are only to first order in g^2, and are not valid if $g^2/\hbar c$ is large.

The author is particularly indebted to Professor H. A. Bethe for his explanation of a method of obtaining finite and gauge invariant results for the problem of vacuum polarization. He is also grateful for Professor Bethe's criticisms of the manuscript, and for innumerable discussions during the development of this work. He wishes to thank Professor J. Ashkin for his careful reading of the manuscript.

APPENDIX

In this appendix a method will be illustrated by which the simpler integrals appearing in problems in electrodynamics can be directly evaluated. The integrals arising in more complex processes lead to rather complicated functions, but the study of the relations of one integral to another and their expression in terms of simpler integrals may be facilitated by the methods given here.

---

[27] The vector meson field potentials φ_μ satisfy

$$-\partial/\partial x_\nu(\partial\varphi_\mu/\partial x_\nu - \partial\varphi_\nu/\partial x_\mu) - \mu^2\varphi_\mu = -4\pi s_\mu,$$

where s_μ, the source for such mesons, is the matrix element of γ_μ between states of neutron and proton. By taking the divergence $\partial/\partial x_\mu$ of both sides, conclude that $\partial\varphi_\nu/\partial x_\nu = 4\pi\mu^{-2}\partial s_\nu/\partial x_\nu$ so that the original equation can be rewritten as

$$\Box^2\varphi_\mu - \mu^2\varphi_\mu = -4\pi(s_\mu + \mu^{-2}\partial/\partial x_\mu(\partial s_\nu/\partial x_\nu)).$$

The right hand side gives in momentum representation $\gamma_\mu - \mu^{-2}q_\mu q_\nu\gamma_\nu$ the left yields the $(q^2-\mu^2)^{-1}$ and finally the interaction $s_\mu\varphi_\mu$ in the Lagrangian gives the γ_μ on absorption.

Proceeding in this way find generally that particles of spin one can be represented by a four-vector u_μ (which, for a free particle of momentum q satisfies $q\cdot u=0$). The propagation of virtual particles of momentum q from state ν to μ is represented by multiplication by the 4–4 matrix (or tensor) $P_{\mu\nu} = (\delta_{\mu\nu} - \mu^{-2}q_\mu q_\nu) \times (q^2-\mu^2)^{-1}$. The first-order interaction (from the Proca equation) with an electromagnetic potential $a\exp(-ik\cdot x)$ corresponds to multiplication by the matrix $E_{\mu\nu} = (q_2\cdot a + q_1\cdot a)\delta_{\mu\nu} - q_{2\nu}a_\mu - q_{1\mu}a_\nu$ where q_1 and $q_2 = q_1 + k$ are the momenta before and after the interaction. Finally, two potentials a, b may act simultaneously, with matrix $E'_{\mu\nu} = -(a\cdot b)\delta_{\mu\nu} + b_\mu a_\nu$.

As a typical problem consider the integral (12) appearing in the first order radiationless scattering problem:

$$\int \gamma_\mu(p_2-k-m)^{-1}a(p_1-k-m)^{-1}\gamma_\mu k^{-2}d^4kC(k^2), \tag{1a}$$

where we shall take $C(k^2)$ to be typically $-\lambda^2(k^2-\lambda^2)^{-1}$ and d^4k means $(2\pi)^{-2}dk_1dk_2dk_3dk_4$. We first rationalize the factors $(p-k-m)^{-1}=(p-k+m)((p-k)^2-m^2)^{-1}$ obtaining,

$$\int \gamma_\mu(p_2-k+m)a(p_1-k+m)\gamma_\mu k^{-2}d^4kC(k^2) \times((p_1-k)^2-m^2)^{-1}((p_2-k)^2-m^2)^{-1}. \tag{2a}$$

The matrix expression may be simplified. It appears to be best to do so *after* the integrations are performed. Since $AB=2A\cdot B-BA$ where $A\cdot B=A_\mu B_\mu$ is a number commuting with all matrices, find, if R is any expression, and A a vector, since $\gamma_\mu A=-A\gamma_\mu+2A_\mu$,

$$\gamma_\mu AR\gamma_\mu=-A\gamma_\mu R\gamma_\mu+2RA. \tag{3a}$$

Expressions between two γ_μ's can be thereby reduced by induction. Particularly useful are

$$\begin{aligned}\gamma_\mu\gamma_\mu&=4\\ \gamma_\mu A\gamma_\mu&=-2A\\ \gamma_\mu AB\gamma_\mu&=2(AB+BA)=4A\cdot B\\ \gamma_\mu ABC\gamma_\mu&=-2CBA\end{aligned} \tag{4a}$$

where A, B, C are any three vector-matrices (i.e., linear combinations of the four γ's).

In order to calculate the integral in (2a) the integral may be written as the sum of three terms (since $k=k_\sigma\gamma_\sigma$),

$$\gamma_\mu(p_2+m)a(p_1+m)\gamma_\mu J_1-[\gamma_\mu\gamma_\sigma a(p_1+m)\gamma_\mu +\gamma_\mu(p_2+m)a\gamma_\sigma\gamma_\mu]J_2+\gamma_\mu\gamma_\sigma a\gamma_\tau\gamma_\mu J_3, \tag{5a}$$

where

$$J_{(1;2;3)}=\int(1;k_\sigma;k_\sigma k_\tau)k^{-2}d^4kC(k^2) \times((p_2-k)^2-m^2)^{-1}((p_1-k)^2-m^2)^{-1}. \tag{6a}$$

That is for J_1 the $(1;k_\sigma;k_\sigma k_\tau)$ is replaced by 1, for J_2 by k_σ, and for J_3 by $k_\sigma k_\tau$.

More complex processes of the first order involve more factors like $((p_3-k)^2-m^2)^{-1}$ and a corresponding increase in the number of k's which may appear in the numerator, as $k_\sigma k_\tau k_\nu\cdots$. Higher order processes involving two or more virtual quanta involve similar integrals but with factors possibly involving $k+k'$ instead of just k, and the integral extending on $k^{-2}d^4kC(k^2)k'^{-2}d^4k'C(k'^2)$. They can be simplified by methods analogous to those used on the first order integrals.

The factors $(p-k)^2-m^2$ may be written

$$(p-k)^2-m^2=k^2-2p\cdot k-\Delta, \tag{7a}$$

where $\Delta=m^2-p^2$, $\Delta_1=m_1^2-p_1^2$, etc., and we can consider dealing with cases of greater generality in that the different denominators need not have the same value of the mass m. In our specific problem (6a), $p_1^2=m^2$ so that $\Delta_1=0$, but we desire to work with greater generality.

Now for the factor $C(k^2)/k^2$ we shall use $-\lambda^2(k^2-\lambda^2)^{-1}k^{-2}$. This can be written as

$$-\lambda^2/(k^2-\lambda^2)k^2=k^{-2}C(k^2)=-\int_0^{\lambda^2}dL(k^2-L)^{-2}. \tag{8a}$$

Thus we can replace $k^{-2}C(k^2)$ by $(k^2-L)^{-2}$ and at the end integrate the result with respect to L from zero to λ^2. We can for many practical purposes consider λ^2 very large relative to m^2 or p^2. When the original integral converges even without the convergence factor, it will be obvious since the L integration will then be convergent to infinity. If an infra-red catastrophe exists in the integral one can simply assume quanta have a small mass $\lambda_{\min}$ and extend the integral on L from $\lambda^2_{\min}$ to λ^2, rather than from zero to λ^2.

We then have to do integrals of the form

$$\int(1;k_\sigma;k_\sigma k_\tau)d^4k(k^2-L)^{-2}(k^2-2p_1\cdot k-\Delta_1)^{-1} \times(k^2-2p_2\cdot k-\Delta_2)^{-1}, \tag{9a}$$

where by $(1;k_\sigma;k_\sigma k_\tau)$ we mean that in the place of this symbol either 1, or k_σ, or $k_\sigma k_\tau$ may stand in different cases. In more complicated problems there may be more factors $(k^2-2p_i\cdot k-\Delta_i)^{-1}$ or other powers of these factors (the $(k^2-L)^{-2}$ may be considered as a special case of such a factor with $p_i=0$, $\Delta_i=L$) and further factors like $k_\sigma k_\tau k_\rho\cdots$ in the numerator. The poles in all the factors are made definite by the assumption that L, and the Δ's have infinitesimal negative imaginary parts.

We shall do the integrals of successive complexity by induction. We start with the simplest convergent one, and show

$$\int d^4k(k^2-L)^{-3}=(8iL)^{-1}. \tag{10a}$$

For this integral is $\int(2\pi)^{-2}dk_4d^3\mathbf{K}(k_4^2-\mathbf{K}\cdot\mathbf{K}-L)^{-3}$ where the vector $\mathbf{K}$, of magnitude $K=(\mathbf{K}\cdot\mathbf{K})^{1/2}$ is k_1, k_2, k_3. The integral on k_4 shows third order poles at $k_4=+(K^2+L)^{1/2}$ and $k_4=-(K^2+L)^{1/2}$. Imagining, in accordance with our definitions, that L has a small negative imaginary part only the first is below the real axis. The contour can be closed by an infinite semi-circle below this axis, without change of the value of the integral since the contribution from the semi-circle vanishes in the limit. Thus the contour can be shrunk about the pole $k_4=+(K^2+L)^{1/2}$ and the resulting k_4 integral is $-2\pi i$ times the residue at this pole. Writing $k_4=(K^2+L)^{1/2}+\epsilon$ and expanding $(k_4^2-K^2-L)^{-3}=\epsilon^{-3}(\epsilon+2(K^2+L)^{1/2})^{-3}$ in powers of ϵ, the residue, being the coefficient of the term ϵ^{-1}, is seen to be $6(2(K^2+L)^{1/2})^{-5}$ so our integral is

$$-(3i/32\pi)\int_0^\infty 4\pi K^2dK(K^2+L)^{-5/2}=(3/8i)(1/3L)$$

establishing (10a).

We also have $\int k_\sigma d^4k(k^2-L)^{-3}=0$ from the symmetry in the k space. We write these results as

$$(8i)\int(1;k_\sigma)d^4k(k^2-L)^{-3}=(1;0)L^{-1}, \tag{11a}$$

where in the brackets $(1;k_\sigma)$ and $(1;0)$ corresponding entries are to be used.

Substituting $k=k'-p$ in (11a), and calling $L-p^2=\Delta$ shows that

$$(8i)\int(1;k_\sigma)d^4k(k^2-2p\cdot k-\Delta)^{-3}=(1;p_\sigma)(p^2+\Delta)^{-1}. \tag{12a}$$

By differentiating both sides of (12a) with respect to Δ, or with respect to p_τ there follows directly

$$(24i)\int(1;k_\sigma;k_\sigma k_\tau)d^4k(k^2-2p\cdot k-\Delta)^{-4} =-(1;p_\sigma;p_\sigma p_\tau-\tfrac{1}{2}\delta_{\sigma\tau}(p^2+\Delta))(p^2+\Delta)^{-2}. \tag{13a}$$

Further differentiations give directly successive integrals including more k factors in the numerator and higher powers of $(k^2-2p\cdot k-\Delta)$ in the denominator.

The integrals so far only contain one factor in the denominator. To obtain results for two factors we make use of the identity

$$a^{-1}b^{-1}=\int_0^1 dx(ax+b(1-x))^{-2}, \tag{14a}$$

(suggested by some work of Schwinger's involving Gaussian integrals). This represents the product of two reciprocals as a parametric integral over one and will therefore permit integrals with two factors to be expressed in terms of one. For other powers of a, b, we make use of all of the identities, such as

$$a^{-2}b^{-1}=\int_0^1 2xdx(ax+b(1-x))^{-3}, \tag{15a}$$

deducible from (14a) by successive differentiations with respect to a or b.

To perform an integral, such as

$$(8i)\int(1;k_\sigma)d^4k(k^2-2p_1\cdot k-\Delta_1)^{-2}(k^2-2p_2\cdot k-\Delta_2)^{-1}, \tag{16a}$$

write, using (15a),

$$(k^2-2p_1\cdot k-\Delta_1)^{-2}(k^2-2p_2\cdot k-\Delta_2)^{-1}=\int_0^1 2xdx(k^2-2p_x\cdot k-\Delta_x)^{-3},$$

where

$$p_x=xp_1+(1-x)p_2 \quad\text{and}\quad \Delta_x=x\Delta_1+(1-x)\Delta_2, \tag{17a}$$

(note that Δ_x is *not* equal to $m^2-p_x^2$) so that the expression (16a) is $(8i)\int_0^1 2xdx\int(1;k_\sigma)d^4k(k^2-2p_x\cdot k-\Delta_x)^{-3}$ which may now be evaluated by (12a) and is

$$(16a)=\int_0^1(1;p_{x\sigma})2xdx(p_x^2+\Delta_x)^{-1}, \tag{18a}$$

where p_x, Δ_x are given in (17a). The integral in (18a) is elementary, being the integral of ratio of polynomials, the denominator of second degree in x. The general expression although readily obtained is a rather complicated combination of roots and logarithms.

Other integrals can be obtained again by parametric differentiation. For example differentiation of (16a), (18a) with respect to Δ_2 or $p_{2\tau}$ gives

$$(8i)\int(1;k_\sigma;k_\sigma k_\tau)d^4k(k^2-2p_1\cdot k-\Delta_1)^{-2}(k^2-2p_2\cdot k-\Delta_2)^{-2}$$
$$=-\int_0^1(1;p_{x\sigma};p_{x\sigma}p_{x\tau}-\tfrac{1}{2}\delta_{\sigma\tau}(p_x^2+\Delta_x))\times 2x(1-x)dx(p_x^2+\Delta_x)^{-2}, \tag{19a}$$

again leading to elementary integrals.

As an example, consider the case that the second factor is just $(k^2-L)^{-2}$ and in the first put $p_1=p$, $\Delta_1=\Delta$. Then $p_x=xp$, $\Delta_x=x\Delta+(1-x)L$. There results

$$(8i)\int(1;k_\sigma;k_\sigma k_\tau)d^4k(k^2-L)^{-2}(k^2-2p\cdot k-\Delta)^{-2}$$
$$=-\int_0^1(1;xp_\sigma;x^2p_\sigma p_\tau-\tfrac{1}{2}\delta_{\sigma\tau}(x^2p^2+\Delta_x))\times 2x(1-x)dx(x^2p^2+\Delta_x)^{-2}. \tag{20a}$$

Integrals with three factors can be reduced to those involving two by using (14a) again. They, therefore, lead to integrals with two parameters (e.g., see application to radiative correction to scattering below).

The methods of calculation given in this paper are deceptively simple when applied to the lower order processes. For processes of increasingly higher orders the complexity and difficulty increases rapidly, and these methods soon become impractical in their present form.

A. Self-Energy

The self-energy integral (19) is

$$(e^2/\pi i)\int\gamma_\mu(p-k-m)^{-1}\gamma_\mu k^{-2}d^4kC(k^2), \tag{19}$$

so that it requires that we find (using the principle of (8a)) the integral on L from 0 to λ^2 of

$$\int\gamma_\mu(p-k+m)\gamma_\mu d^4k(k^2-L)^{-2}(k^2-2p\cdot k)^{-1},$$

since $(p-k)^2-m^2=k^2-2p\cdot k$, as $p^2=m^2$. This is of the form (16a) with $\Delta_1=L$, $p_1=0$, $\Delta_2=0$, $p_2=p$ so that (18a) gives, since $p_x=(1-x)p$, $\Delta_x=xL$,

$$(8i)\int(1;k_\sigma)d^4k(k^2-L)^{-2}(k^2-2p\cdot k)^{-1}$$
$$=\int_0^1(1;(1-x)p_\sigma)2xdx((1-x)^2m^2+xL)^{-1},$$

or performing the integral on L, as in (8),

$$(8i)\int(1;k_\sigma)d^4kk^{-2}C(k^2)(k^2-2p\cdot k)^{-1}$$
$$=\int_0^1(1;(1-x)p_\sigma)2dx\ln\frac{x\lambda^2+(1-x)^2m^2}{(1-x)^2m^2}.$$

Assuming now that $\lambda^2\gg m^2$ we neglect $(1-x)^2m^2$ relative to $x\lambda^2$ in the argument of the logarithm, which then becomes $(\lambda^2/m^2)(x/(1-x)^2)$. Then since $\int_0^1dx\ln(x(1-x)^{-2})=1$ and $\int_0^1(1-x)dx\ln(x(1-x)^{-2})=-(1/4)$ find

$$(8i)\int(1;k_\sigma)k^{-2}C(k^2)d^4k(k^2-2p\cdot k)^{-1}$$
$$=\left(2\ln\frac{\lambda^2}{m^2}+2;\,p_\sigma\left(\ln\frac{\lambda^2}{m^2}-\frac{1}{2}\right)\right),$$

so that substitution into (19) (after the $(p-k-m)^{-1}$ in (19) is replaced by $(p-k+m)(k^2-2p\cdot k)^{-1}$) gives

$$\begin{aligned}(19)&=(e^2/8\pi)\gamma_\mu[(p+m)(2\ln(\lambda^2/m^2)+2)-p(\ln(\lambda^2/m^2)-\tfrac{1}{2})]\gamma_\mu\\&=(e^2/8\pi)[8m(\ln(\lambda^2/m^2)+1)-p(2\ln(\lambda^2/m^2)+5)],\end{aligned} \tag{20}$$

using (4a) to remove the γ_μ's. This agrees with Eq. (20) of the text, and gives the self-energy (21) when p is replaced by m.

B. Corrections to Scattering

The term (12) in the radiationless scattering, after rationalizing the matrix denominators and using $p_1^2=p_2^2=m^2$ requires the integrals (9a), as we have discussed. This is an integral with three denominators which we do in two stages. First the factors $(k^2-2p_1\cdot k)$ and $(k^2-2p_2\cdot k)$ are combined by a parameter y;

$$(k^2-2p_1\cdot k)^{-1}(k^2-2p_2\cdot k)^{-1}=\int_0^1dy(k^2-2p_y\cdot k)^{-2},$$

from (14a) where

$$p_y=yp_1+(1-y)p_2. \tag{21a}$$

We therefore need the integrals

$$(8i)\int(1;k_\sigma;k_\sigma k_\tau)d^4k(k^2-L)^{-2}(k^2-2p_y\cdot k)^{-2}, \tag{22a}$$

which we will then integrate with respect to y from 0 to 1. Next we do the integrals (22a) immediately from (20a) with $p=p_y$, $\Delta=0$:

$$(22a)=-\int_0^1\int_0^1(1;xp_{y\sigma};x^2p_{y\sigma}p_{y\tau}-\tfrac{1}{2}\delta_{\sigma\tau}(x^2p_y^2+(1-x)L))2x(1-x)dx(x^2p_y^2+L(1-x))^{-2}dy.$$

We now turn to the integrals on L as required in (8a). The first term, (1), in $(1;k_\sigma;k_\sigma k_\tau)$ gives no trouble for large L, but if L is put equal to zero there results $x^{-2}p_y^{-2}$ which leads to a diverging integral on x as $x\rightarrow 0$. This infra-red catastrophe is analyzed by using $\lambda_{\min}^2$ for the lower limit of the L integral. For the last term the upper limit of L must be kept as λ^2. Assuming $\lambda_{\min}^2\ll p_y^2\ll\lambda^2$ the x integrals which remain are trivial, as in the self-energy case. One finds

$$-(8i)\int(k^2-\lambda_{\min}^2)^{-1}d^4kC(k^2-\lambda_{\min}^2)(k^2-2p_1\cdot k)^{-1}(k^2-2p_2\cdot k)^{-1}$$
$$=\int_0^1p_y^{-2}dy\ln(p_y^2/\lambda_{\min}^2) \tag{23a}$$

$$-(8i)\int k_\sigma k^{-2}d^4kC(k^2)(k^2-2p_1\cdot k)^{-1}(k^2-2p_2\cdot k)^{-1}$$
$$=2\int_0^1p_{y\sigma}p_y^{-2}dy, \tag{24a}$$

$$-(8i)\int k_\sigma k_\tau k^{-2}d^4kC(k^2)(k^2-2p_1\cdot k)^{-1}(k^2-2p_2\cdot k)^{-1}$$
$$=\int_0^1p_{y\sigma}p_{y\tau}p_y^{-2}dy-\tfrac{1}{2}\delta_{\sigma\tau}\int_0^1dy\ln(\lambda^2p_y^{-2})+\tfrac{1}{4}\delta_{\sigma\tau}. \tag{25a}$$

The integrals on y give,

$$\int_0^1p_y^{-2}dy\ln(p_y^2\lambda_{\min}^{-2})=4(m^2\sin2\theta)^{-1}\left[\theta\ln(m\lambda_{\min}^{-1})-\int_0^\theta\alpha\tan\alpha d\alpha\right], \tag{26a}$$

$$\int_0^1p_{y\sigma}p_y^{-2}dy=\theta(m^2\sin2\theta)^{-1}(p_{1\sigma}+p_{2\sigma}), \tag{27a}$$

$$\int_0^1p_{y\sigma}p_{y\tau}p_y^{-2}dy=\theta(2m^2\sin2\theta)^{-1}(p_{1\sigma}+p_{1\tau})(p_{2\sigma}+p_{2\tau})+q^{-2}q_\sigma q_\tau(1-\theta\,\text{ctn}\theta), \tag{28a}$$

$$\int_0^1dy\ln(\lambda^2p_y^{-2})=\ln(\lambda^2/m^2)+2(1-\theta\,\text{ctn}\theta). \tag{29a}$$

These integrals on y were performed as follows. Since $p_2=p_1+q$ where q is the momentum carried by the potential, it follows from $p_2^2=p_1^2=m^2$ that $2p_1\cdot q=-q^2$ so that since $p_y=p_1+q(1-y)$, $p_y^2=m^2-q^2y(1-y)$. The substitution $2y-1=\tan\alpha/\tan\theta$ where θ is defined by $4m^2\sin^2\theta=q^2$ is useful for it means $p_y^2=m^2\sec^2\alpha/\sec^2\theta$ and $p_y^{-2}dy=(m^2\sin 2\theta)^{-1}d\alpha$ where α goes from $-\theta$ to $+\theta$.

These results are substituted into the original scattering formula (2a), giving (22). It has been simplified by frequent use of the fact that p_1 operating on the initial state is m, and likewise p_2 when it appears at the left is replacable by m. (Thus, to simplify:

$$\begin{aligned}\gamma_\mu p_2 a p_1 \gamma_\mu &= -2p_1 a p_2 \text{ by (4a)},\\ &= -2(p_2-q)a(p_1+q) = -2(m-q)a(m+q).\end{aligned}$$

A term like $qaq=-q^2a+2(a\cdot q)q$ is equivalent to just $-q^2a$ since $q=p_2-p_1=m-m$ has zero matrix element.) The renormalization term requires the corresponding integrals for the special case $q=0$.

C. Vacuum Polarization

The expressions (32) and (32′) for $J_{\mu\nu}$ in the vacuum polarization problem require the calculation of the integral

$$J_{\mu\nu}(m^2)=-\frac{e^2}{\pi i}\int Sp[\gamma_\mu(p-\tfrac{1}{2}q+m)\gamma_\nu(p+\tfrac{1}{2}q+m)]d^4p \times((p-\tfrac{1}{2}q)^2-m^2)^{-1}((p+\tfrac{1}{2}q)^2-m^2)^{-1}, \quad (32)$$

where we have replaced p by $p-\frac{1}{2}q$ to simplify the calculation somewhat. We shall indicate the method of calculation by studying the integral,

$$I(m^2)=\int p_\sigma p_\tau d^4p((p-\tfrac{1}{2}q)^2-m^2)^{-1}((p+\tfrac{1}{2}q)^2-m^2)^{-1}.$$

The factors in the denominator, $p^2-p\cdot q-m^2+\frac{1}{4}q^2$ and $p^2+p\cdot q-m^2+\frac{1}{4}q^2$ are combined as usual by (8a) but for symmetry we substitute $x=\frac{1}{2}(1+\eta)$, $(1-x)=\frac{1}{2}(1-\eta)$ and integrate η from -1 to $+1$:

$$I(m^2)=\int_{-1}^{+1} p_\sigma p_\tau d^4p(p^2-\eta p\cdot q-m^2+\tfrac{1}{4}q^2)^{-2}d\eta/2. \quad (30a)$$

But the integral on p will not be found in our list for it is badly divergent. However, as discussed in Section 7, Eq. (32′) we do not wish $I(m^2)$ but rather $\int_0^\infty[I(m^2)-I(m^2+\lambda^2)]G(\lambda)d\lambda$. We can calculate the difference $I(m^2)-I(m^2+\lambda^2)$ by first calculating the derivative $I'(m^2+L)$ of I with respect to m^2 at m^2+L and later integrating L from zero to λ^2. By differentiating (30a), with respect to m^2 find,

$$I'(m^2+L)=\int_{-1}^{+1} p_\sigma p_\tau d^4p(p^2-\eta p\cdot q-m^2-L+\tfrac{1}{4}q^2)^{-3}d\eta.$$

This still diverges, but we can differentiate again to get

$$\begin{aligned}I''(m^2+L)&=3\int_{-1}^{+1} p_\sigma p_\tau d^4p(p^2-\eta p\cdot q-m^2-L+\tfrac{1}{4}q^2)^{-4}d\eta\\ &=-(8i)^{-1}\int_{-1}^{+1}(\tfrac{1}{4}\eta^2 q_\sigma q_\tau D^{-2}-\tfrac{1}{2}\delta_{\sigma\tau}D^{-1})d\eta\end{aligned} \quad (31a)$$

(where $D=\frac{1}{4}(\eta^2-1)q^2+m^2+L$), which now converges and has been evaluated by (13a) with $p=\frac{1}{2}\eta q$ and $\Delta=m^2+L-\frac{1}{4}q^2$. Now to get I' we may integrate I'' with respect to L as an indefinite integral and *we may choose any convenient arbitrary constant*. This is because a constant C in I' will mean a term $-C\lambda^2$ in $I(m^2)-I(m^2+\lambda^2)$ which vanishes since we will integrate the results times $G(\lambda)d\lambda$ and $\int_0^\infty\lambda^2G(\lambda)d\lambda=0$. This means that the logarithm appearing on integrating L in (31a) presents no problem. We may take

$$I'(m^2+L)=(8i)^{-1}\int_{-1}^{+1}[\tfrac{1}{4}\eta^2q_\sigma q_\tau D^{-1}+\tfrac{1}{2}\delta_{\sigma\tau}\ln D]d\eta+C\delta_{\sigma\tau},$$

a subsequent integral on L and finally on η presents no new problems. There results

$$\begin{aligned}&-(8i)\int p_\sigma p_\tau d^4p((p-\tfrac{1}{2}q)^2-m^2)^{-1}((p+\tfrac{1}{2}q)^2-m^2)^{-1}\\ &\quad=(q_\sigma q_\tau-\delta_{\sigma\tau}q^2)\left[\frac{1}{9}-\frac{4m^2-q^2}{3q^2}\left(1-\frac{\theta}{\tan\theta}\right)+\frac{1}{6}\ln\frac{\lambda^2}{m^2}\right]\\ &\quad\quad+\delta_{\sigma\tau}[(\lambda^2+m^2)\ln(\lambda^2m^{-2}+1)-C'\lambda^2],\end{aligned} \quad (32a)$$

where we assume $\lambda^2 \gg m^2$ and have put some terms into the arbitrary constant C' which is independent of λ^2 (but in principle could depend on q^2) and which drops out in the integral on $G(\lambda)d\lambda$. We have set $q^2=4m^2\sin^2\theta$.

In a very similar way the integral with m^2 in the numerator can be worked out. It is, of course, necessary to differentiate this m^2 also when calculating I' and I''. There results

$$\begin{aligned}&-(8i)\int m^2d^4p((p-\tfrac{1}{2}q)^2-m^2)^{-1}((p+\tfrac{1}{2}q)^2-m^2)^{-1}\\ &=4m^2(1-\theta\,\mathrm{ctn}\,\theta)-q^2/3+2(\lambda^2+m^2)\ln(\lambda^2m^{-2}+1)-C''\lambda^2),\end{aligned} \quad (33a)$$

with another unimportant constant C''. The complete problem requires the further integral,

$$\begin{aligned}&-(8i)\int(1;p_\sigma)d^4p((p-\tfrac{1}{2}q)^2-m^2)^{-1}((p+\tfrac{1}{2}q)^2-m^2)^{-1}\\ &\quad=(1,0)(4(1-\theta\,\mathrm{ctn}\,\theta)+2\ln(\lambda^2m^{-2})).\end{aligned} \quad (34a)$$

The value of the integral (34a) times m^2 differs from (33a), of course, because the results on the right are not actually the integrals on the left, but rather equal their actual value minus their value for $m^2=m^2+\lambda^2$.

Combining these quantities, as required by (32), dropping the constants C', C'' and evaluating the spur gives (33). The spurs are evaluated in the usual way, noting that the spur of any odd number of γ matrices vanishes and $Sp(AB)=Sp(BA)$ for arbitrary A, B. The $Sp(1)=4$ and we also have

$$\tfrac{1}{4}Sp[(p_1+m_1)(p_2-m_2)]=p_1\cdot p_2-m_1m_2, \quad (35a)$$

$$\begin{aligned}&\tfrac{1}{4}Sp[(p_1+m_1)(p_2-m_2)(p_3+m_3)(p_4-m_4)]\\ &\quad=(p_1\cdot p_2-m_1m_2)(p_3\cdot p_4-m_3m_4)\\ &\quad\quad-(p_1\cdot p_3-m_1m_3)(p_2\cdot p_4-m_2m_4)\\ &\quad\quad+(p_1\cdot p_4-m_1m_4)(p_2\cdot p_3-m_2m_3),\end{aligned} \quad (36a)$$

where p_i, m_i are arbitrary four-vectors and constants.

It is interesting that the terms of order $\lambda^2\ln\lambda^2$ go out, so that the charge renormalization depends only logarithmically on λ^2. This is not true for some of the meson theories. Electrodynamics is suspiciously unique in the mildness of its divergence.

D. More Complex Problems

Matrix elements for complex problems can be set up in a manner analogous to that used for the simpler cases. We give three illustrations; higher order corrections to the Møller scatter-

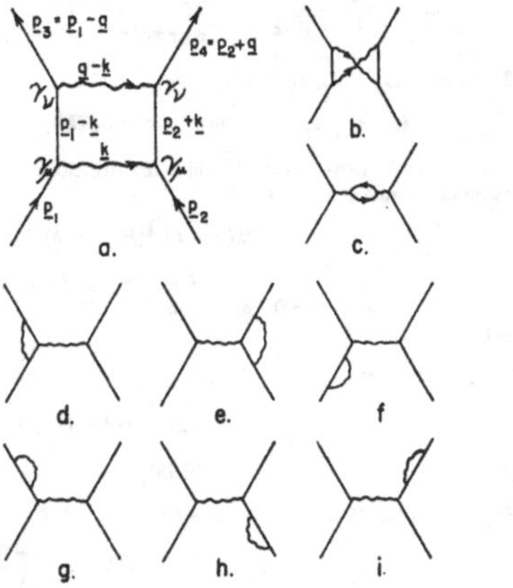

FIG. 8. The interaction between two electrons to order $(e^2/\hbar c)^2$. One adds the contribution of every figure involving two virtual quanta, Appendix D.

ing, to the Compton scattering, and the interaction of a neutron with an electromagnetic field.

For the Møller scattering, consider two electrons, one in state u_1 of momentum p_1 and the other in state u_2 of momentum p_2. Later they are found in states u_3, p_3 and u_4, p_4. This may happen (first order in $e^2/\hbar c$) because they exchange a quantum of momentum $q=p_1-p_3=p_4-p_2$ in the manner of Eq. (4) and Fig. 1. The matrix element for this process is proportional to (translating (4) to momentum space)

$$(\bar{u}_4\gamma_\mu u_2)(\bar{u}_3\gamma_\mu u_1)q^{-2}. \quad (37a)$$

We shall discuss corrections to (37a) to the next order in $e^2/\hbar c$. (There is also the possibility that it is the electron at 2 which finally arrives at 3, the electron at 1 going to 4 through the exchange of quantum of momentum p_3-p_2. The amplitude for this process, $(\bar{u}_4\gamma_\mu u_1)(\bar{u}_3\gamma_\mu u_2)(p_3-p_2)^{-2}$, must be subtracted from (37a) in accordance with the exclusion principle. A similar situation exists to each order so that we need consider in detail only the corrections to (37a), reserving to the last the subtraction of the same terms with 3, 4 exchanged.)

One reason that (37a) is modified is that two quanta may be exchanged, in the manner of Fig. 8a. The total matrix element for all exchanges of this type is

$$(e^2/\pi i)\int(\bar{u}_3\gamma_\nu(p_1-k-m)^{-1}\gamma_\mu u_1)(\bar{u}_4\gamma_\nu(p_2+k-m)^{-1}\gamma_\mu u_2) \cdot k^{-2}(q-k)^{-2}d^4k, \quad (38a)$$

as is clear from the figure and the general rule that electrons of momentum p contribute in amplitude $(p-m)^{-1}$ between interactions γ_μ, and that quanta of momentum k contribute k^{-2}. In integrating on d^4k and summing over μ and ν. we add all alternatives of the type of Fig. 8a. If the time of absorption, γ_μ, of the quantum k by electron 2 is later than the absorption, γ_ν, of $q-k$, this corresponds to the virtual state p_2+k being a positron (so that (38a) contains over thirty terms of the conventional method of analysis).

In integrating over all these alternatives we have considered all possible distortions of Fig. 8a which preserve the order of events along the trajectories. We have not included the possibilities corresponding to Fig. 8b, however. Their contribution is

$$(e^2/\pi i)\int(\bar{u}_3\gamma_\nu(p_1-k-m)^{-1}\gamma_\mu u_1) \times(\bar{u}_4\gamma_\mu(p_2+q-k-m)^{-1}\gamma_\nu u_2)k^{-2}(q-k)^{-2}d^4k, \quad (39a)$$

as is readily verified by labeling the diagram. The contributions of all possible ways that an event can occur are to be added. This means that one adds with equal weight the integrals corresponding to each topologically distinct figure.

To this same order there are also the possibilities of Fig. 8d which give

$$(e^2/\pi i)\int(\bar{u}_3\gamma_\nu(p_3-k-m)^{-1}\gamma_\mu(p_1-k-m)^{-1}\gamma_\nu u_1) \times(\bar{u}_4\gamma_\mu u_2)k^{-2}q^{-2}d^4k.$$

This integral on k will be seen to be precisely the integral (12) for the radiative corrections to scattering, which we have worked out. The term may be combined with the renormalization terms resulting from the difference of the effects of mass change and the terms, Figs. 8f and 8g. Figures 8e, 8h, and 8i are similarly analyzed.

Finally the term Fig. 8c is clearly related to our vacuum polarization problem, and when integrated gives a term proportional to $(\bar{u}_4\gamma_\mu u_2)(\bar{u}_3\gamma_\nu u_1)J_{\mu\nu}q^{-4}$. If the charge is renormalized the term $\ln(\lambda/m)$ in $J_{\mu\nu}$ in (33) is omitted so there is no remaining dependence on the cut-off.

The only new integrals we require are the convergent integrals (38a) and (39a). They can be simplified by rationalizing the denominators and combining them by (14a). For example (38a) involves the factors $(k^2-2p_1\cdot k)^{-1}(k^2+2p_2\cdot k)^{-1}k^{-2}(q^2+k^2-2q\cdot k)^{-2}$. The first two may be combined by (14a) with a parameter x, and the second pair by an expression obtained by differentiation (15a) with respect to b and calling the parameter y. There results a factor $(k^2-2p_x\cdot k)^{-2}(k^2+yq^2-2yq\cdot k)^{-4}$ so that the integrals on d^4k now involve two factors and can be performed by the methods given earlier in the appendix. The subsequent integrals on the parameters x and y are complicated and have not been worked out in detail.

Working with charged mesons there is often a considerable reduction of the number of terms. For example, for the interaction between protons resulting from the exchange of two mesons only the term corresponding to Fig. 8b remains. Term 8a, for example, is impossible, for if the first proton emits a positive meson the second cannot absorb it directly for only neutrons can absorb positive mesons.

As a second example, consider the radiative correction to the Compton scattering. As seen from Eq. (15) and Fig. 5 this scattering is represented by two terms, so that we can consider the corrections to each one separately. Figure 9 shows the types of terms arising from corrections to the term of Fig. 5a. Calling k the momentum of the virtual quantum, Fig. 9a gives an integral

$$\int\gamma_\mu(p_2-k-m)^{-1}e_2(p_1+q_1-k-m)^{-1}e_1(p_1-k-m)^{-1}\gamma_\mu k^{-2}d^4k,$$

convergent without cut-off and reducible by the methods outlined in this appendix.

The other terms are relatively easy to evaluate. Terms b and c of Fig. 9 are closely related to radiative corrections (although somewhat more difficult to evaluate, for one of the states is not that of a free electron, $(p_1+q)^2\neq m^2$). Terms e, f are renormalization terms. From term d must be subtracted explicitly the effect of mass Δm, as analyzed in Eqs. (26) and (27) leading to (28) with $p'=p_1+q$, $a=e_2$, $b=e_1$. Terms g, h give zero since the vacuum polarization has zero effect on free light quanta, $q_1^2=0$, $q_2^2=0$. The total is insensitive to the cut-off λ.

The result shows an infra-red catastrophe, the largest part of the effect. When cut-off at λ_{min}, the effect proportional to $\ln(m/\lambda_{min})$ goes as

$$(e^2/\pi)\ln(m/\lambda_{min})(1-2\theta\,\mathrm{ctn}\,2\theta), \quad (40a)$$

times the uncorrected amplitude, where $(p_2-p_1)^2=4m^2\sin^2\theta$. This is the same as for the radiative correction to scattering for a deflection p_2-p_1. This is physically clear since the long wave quanta are not effected by short-lived intermediate states. The infra-red effects arise[28] from a final adjustment of the field from the asymptotic coulomb field characteristic of the electron of

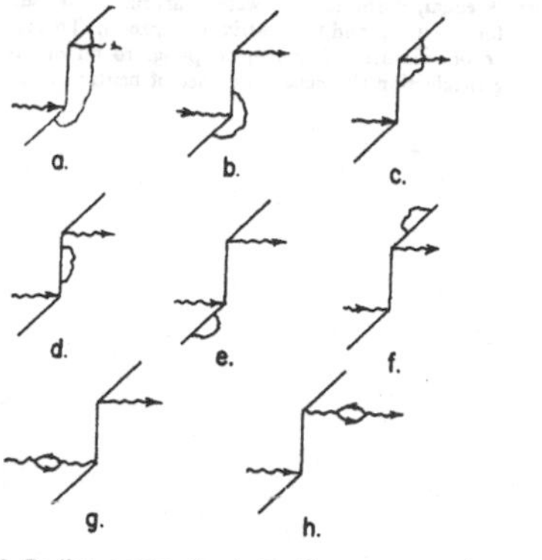

FIG. 9. Radiative correction to the Compton scattering term (a) of Fig. 5. Appendix D.

[28] F. Bloch and A. Nordsieck, Phys. Rev. 52, 54 (1937).

momentum p_1 before the collision to that characteristic of an electron moving in a new direction p_2 after the collision.

The complete expression for the correction is a very complicated expression involving transcendental integrals.

As a final example we consider the interaction of a neutron with an electromagnetic field in virtue of the fact that the neutron may emit a virtual negative meson. We choose the example of pseudoscalar mesons with pseudovector coupling. The change in amplitude due to an electromagnetic field $A=a\exp(-iq\cdot x)$ determines the scattering of a neutron by such a field. In the limit of small q it will vary as $qa-aq$ which represents the interaction of a particle possessing a magnetic moment. The first-order interaction between an electron and a neutron is given by the same calculation by considering the exchange of a quantum between the electron and the nucleon. In this case a_μ is q^{-2} times the matrix element of γ_μ between the initial and final states of the electron, the states differing in momentum by q.

The interaction may occur because the neutron of momentum p_1 emits a negative meson becoming a proton which proton interacts with the field and then reabsorbs the meson (Fig. 10a). The matrix for this process is ($p_2=p_1+q$),

$$\int(\gamma_5 k)(p_2-k-M)^{-1}a(p_1-k-M)^{-1}(\gamma_5 k)(k^2-\mu^2)^{-1}d^4k. \quad (41a)$$

Alternatively it may be the meson which interacts with the field. We assume that it does this in the manner of a scalar potential satisfying the Klein Gordon Eq. (35), (Fig. 10b)

$$-\int(\gamma_5 k_2)(p_1-k_1-M)^{-1}(\gamma_5 k_1)(k_2^2-\mu^2)^{-1} \times(k_2\cdot a+k_1\cdot a)(k_1^2-\mu^2)^{-1}d^4k_1, \quad (42a)$$

where we have put $k_2=k_1+q$. The change in sign arises because the virtual meson is negative. Finally there are two terms arising from the $\gamma_5 a$ part of the pseudovector coupling (Figs. 10c, 10d)

$$\int(\gamma_5 k)(p_2-k-M)^{-1}(\gamma_5 a)(k^2-\mu^2)^{-1}d^4k, \quad (43a)$$

and

$$\int(\gamma_5 a)(p_1-k-M)^{-1}(\gamma_5 k)(k^2-\mu^2)^{-1}d^4k. \quad (44a)$$

Using convergence factors in the manner discussed in the section on meson theories each integral can be evaluated and the results combined. Expanded in powers of q the first term gives the magnetic moment of the neutron and is insensitive to the cut-off, the next gives the scattering amplitude of slow electrons on neutrons, and depends logarithmically on the cut-off.

The expressions may be simplified and combined somewhat before integration. This makes the integrals a little easier and also shows the relation to the case of pseudoscalar coupling. For example in (41a) the final $\gamma_5 k$ can be written as $\gamma_5(k-p_1+M)$ since $p_1=M$ when operating on the initial neutron state. This is

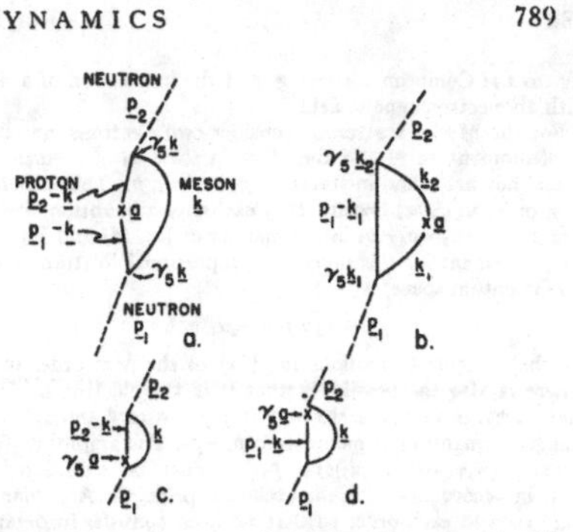

FIG. 10. According to the meson theory a neutron interacts with an electromagnetic potential a by first emitting a virtual charged meson. The figure illustrates the case for a pseudoscalar meson with pseudovector coupling. Appendix D.

$(p_1-k-M)\gamma_5+2M\gamma_5$ since γ_5 anticommutes with p_1 and k. The first term cancels the $(p_1-k-M)^{-1}$ and gives a term which just cancels (43a). In a like manner the leading factor $\gamma_5 k$ in (41a) is written as $-2M\gamma_5-\gamma_5(p_2-k-M)$, the second term leading to a simpler term containing no $(p_2-k-M)^{-1}$ factor and combining with a similar one from (44a). One simplifies the $\gamma_5 k_1$ and $\gamma_5 k_2$ in (42a) in an analogous way. There finally results terms like (41a), (42a) but with pseudoscalar coupling $2M\gamma_5$ instead of $\gamma_5 k$, no terms like (43a) or (44a) and a remainder, representing the difference in effects of pseudovector and pseudoscalar coupling. The pseudoscalar terms do not depend sensitively on the cut-off, but the difference term depends on it logarithmically. The difference term affects the electron-neutron interaction but not the magnetic moment of the neutron.

Interaction of a proton with an electromagnetic potential can be similarly analyzed. There is an effect of virtual mesons on the electromagnetic properties of the proton even in the case that the mesons are neutral. It is analogous to the radiative corrections to the scattering of electrons due to virtual photons. The sum of the magnetic moments of neutron and proton for charged mesons is the same as the proton moment calculated for the corresponding neutral mesons. In fact it is readily seen by comparing diagrams, that for arbitrary q, the scattering matrix to *first order in the electromagnetic potential* for a proton according to neutral meson theory is equal, if the mesons were charged, to the sum of the matrix for a neutron and the matrix for a proton. This is true, for any type or mixtures of meson coupling, to all orders in the coupling (neglecting the mass difference of neutron and proton).

Printed in the United States
by Baker & Taylor Publisher Services